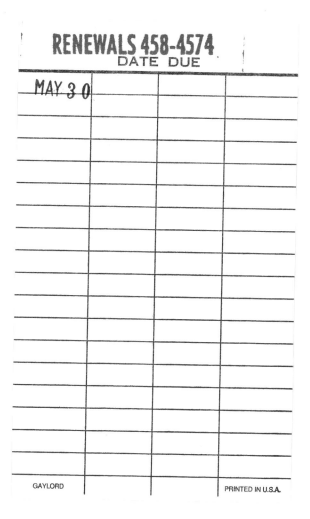

RENEWALS 458-4574

DATE DUE

MAY 3 0			
GAYLORD			PRINTED IN U.S.A.

Gamma

LEONARD EULER.

Leonhard Euler (1707–1783)

Read Euler, read Euler, he is master of us all.
Pierre-Simon Laplace (1749–1827)

Euler calculated without effort, just as men breathe,
as eagles sustain themselves in the air.
Dominique François Jean Arago (1786–1853)

The study of Euler's works remains the best instruction in the various areas of
mathematics and can be replaced by no other.
Carl Frederick Gauss (1777–1855)

Gamma

EXPLORING EULER'S CONSTANT

Julian Havil

PRINCETON UNIVERSITY PRESS

PRINCETON AND OXFORD

Copyright © 2003 by Princeton University Press
Published by Princeton University Press, 41 William Street,
Princeton, New Jersey 08540
In the United Kingdom: Princeton University Press, 3 Market Place,
Woodstock, Oxfordshire OX20 1SY

Library of Congress Cataloguing-in-Publication Data

Havil, Julian, 1952–
Gamma: exploring Euler's constant/Julian Havil
p. cm.
Includes bibliographical references and index.
ISBN 0-691-09983-9 (acid-free paper)
1. Mathematical constants. 2. Euler, Leonhard, 1707–1783. I. Title
QA41.H23   2003
513—dc21   2002192453

British Library Cataloguing-in-Publication Data

A catalogue record for this book is available from the British Library.

This book has been composed in Times. Typeset by T&T Productions Ltd, London.
Printed on acid-free paper. ∞
www.pupress.princeton.edu

Printed in the United States of America

10 9 8 7 6 5 4 3 2 1

I had a feeling once about Mathematics—that I saw it all. Depth beyond depth was revealed to me—the Byss and Abyss. I saw—as one might see the transit of Venus or even the Lord Mayor's Show—a quantity passing through infinity and changing its sign from plus to minus. I saw exactly why it happened and why the tergiversation was inevitable but it was after dinner and I let it go.

Sir Winston Churchill (1874–1965)

Contents

Foreword

I am delighted to be allowed to add a few words to this book by Julian Havil, who is a teacher of mathematics at the school where I was a student sixty years ago. I fell in love with mathematics at the school and have been a professional mathematician ever since.

This book is not for professional mathematicians but rather it is aimed at students of mathematics, be they eager high school students or undergraduates, and those who teach them. It is an inspiring book that will give them an idea of how enchanting mathematics can be.

Mathematics is often thought to be difficult and dull. Many people avoid it as much as they can and as a result much of the population is mathematically illiterate. This is in part due to the relative lack of importance given to numeracy in our culture, and to the way that the subject has been presented to students. It could be argued that the two most widely used approaches to teaching mathematics, at school level and beyond, have themselves contributed to this level of mathematical illiteracy.

The first approach was the 'boot-camp' method of drill and exercise that prepared students well for examinations but often did not enable them to develop a real understanding of mathematics. It mostly failed to encourage students to see the beauty and enjoyment to be gained from the subject. I remember this style well from my school years, where we used the successful and influential textbooks written by our own head of mathematics, Clement Durell.

The second approach, very much in fashion when my own children were at school, was called 'New Math' and was a reaction to the dullness and shallowness of the old way of teaching. The New Math teaching was based on the idea that children should learn to understand modern mathematical concepts before they learned to solve practical problems, hence students would learn about sets and relations before they had mastered multiplication and division. Students learned the vocabulary of modern mathematics without understanding the substance. After a few years of New Math, mathematical literacy declined precipitously.

Is there a third approach that could be more successful? I believe there is a promising third way, and this book by Havil shows us where to find it. The third way is to use a historical approach to mathematics, teaching the practical skills that students need, but in the context of the history of the time when these skills were first developed.

Havil has chosen the 18th century as the context to be studied. This is the right choice. In the 18th century, the tricks and ideas of higher mathematics arose naturally out of the practical problems of the day. The sharp modern divisions of mathematics into pure and applied, abstract and concrete, did not yet exist. The presiding genius of Leonhard Euler created the language and the style in which mathematics has developed ever since. This book is centred on the personality of Euler and the ideas that he left for his successors to use and ponder. Euler's ideas are simple enough to be accessible, and deep enough to give a feeling for the beauty of real mathematics.

In this book, as is so often the case in mathematics, a little effort on the part of the reader will open a world of ideas. The book is so much more than an account of a few subjects within mathematics or a list of examples of Euler's genius. Anyone who has the least inkling that mathematics is important, interesting and beautiful will find the book inspiring, and very enjoyable.

In conclusion I say to the teachers and students who may use this book: Here is a cupboard full of bottles of vintage wine. Now drink!

Freeman Dyson

Acknowledgements

An author of a book needs help and support and a first-time author needs plenty of both. Thanks to my colleagues, Dr John Hodgins and Miss Coralie Ovenden, who respectively looked at an early form of the typescript and translated 18th-century Latin into 21st-century English. To another colleague, and great friend, Lachlan Mackinnon, who showed interest and took time out from his distinguished literary career to help a mathematician write what he wanted to say. To Charlotte Liu, for her contagious enthusiasm for the project. To students Owen Jones and Andrei Pogonaru, who bravely questioned, and by so doing, greatly improved the typescript. To the many friends in my local watering hole, the Wykeham Arms, who have shown considerable interest and enthusiasm for a project so distant from their own worlds. Thanks to them all. Also, to the previewers and reviewers for their time and helpful comments and most particularly to Professor Freeman Dyson for taking the trouble to write the preface when he has so many more important things to do. To the many authors of articles and books whose scholarship contributed to my own; particularly, the Mac Tutor History of Mathematics site for the most useful biographical details and dates. The package Mathematica, which proved so useful in so many ways, and the work of Stan Wagon, who saved the author countless hours of programming by the publication of his Mathematical Explorer software and *Mathematica in Action* book. To Jonathan Wainwright for his patient and informed typesetting, always accepting without demur yet another addition or alteration to the 'finished' typescript. Lastly, thanks to my editor, David Ireland, for his calm, supportive, diplomatic, cheery and professional approach; I cannot believe that all editors are so good.

The poem 'The Riemann Conjecture' on p. 216 is copyright Jonathan P. Dowling, reproduced by permission of the author. The frontispiece portrait of Euler and the picture of Napier's bones on p. 19 are provided courtesy of Ann Ronan, Picture Library, copyright Picture Library.

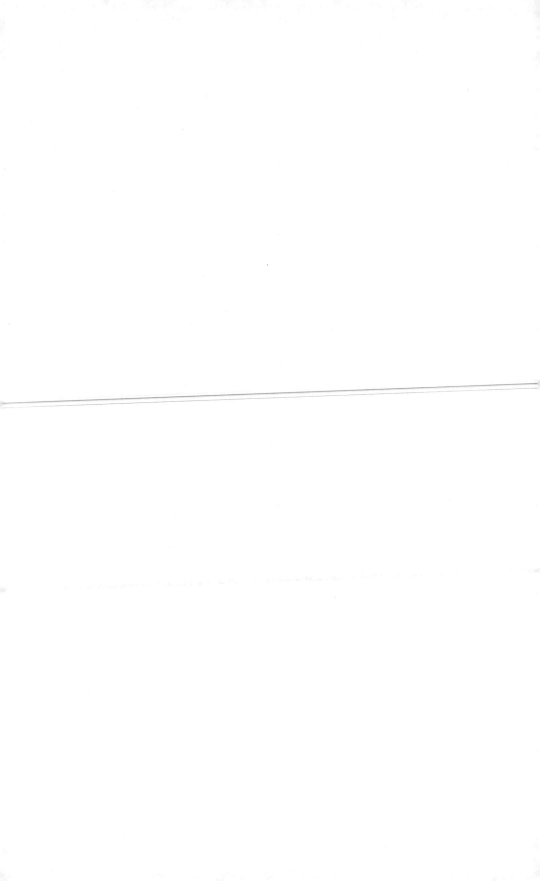

The last thing one knows when writing a book is what to put first.

Blaise Pascal (1623–1662)

It is tempting to think that there are just three special mathematical constants: π, e and i. In fact there are many, each with its own definition, each originating in some natural way in its own area of mathematics, each given a special symbol and a name too. They need symbols to represent them because they are awkward; that is, they have no convenient, finite numeric representation and no patterned infinite one: the ratio of the circumference to the diameter of any circle is not 3.142 or $\frac{22}{7}$, it is 3.141 59..., which is as mysterious as $(2.718\,28\ldots)^x$ essentially being the only function equal to its own derivative; in each case the trailing dots suggest the irrationality (let alone transcendence) of the numbers. Compared with these, writing i for $\sqrt{-1}$ is a small convenience. The number, now universally known as Gamma, is generally accepted to be the most significant of the 'constants obscura' and as such is the fourth important special constant of mathematics; its symbol is the Greek letter γ and the constant it represents is forever associated with the name of the Swiss genius, Leonhard Euler (1707–1783). Its value is the unprepossessing 0.577 215 6..., with its own trailing dots making the same suggestions about its character—but unlike its illustrious colleagues, so far they remain no more than suggestions.

This book is an exploration of γ and inescapably this means that it is also an exploration of logarithms and the harmonic series, since it is the interrelationship between them that Euler exploited to define his constant as

$$\gamma = \lim_{n\to\infty}\left(1 + \frac{1}{2} + \frac{1}{3} + \frac{1}{4} + \cdots + \frac{1}{n} - \ln n\right),$$

where the ln is the ubiquitous log to the base e, derived from the French expression 'logarithmic natural'; the harmonic series, which occupies a less publicized place in mathematical literature, is its discrete counterpart:

$$H_n = 1 + \frac{1}{2} + \frac{1}{3} + \frac{1}{4} + \cdots + \frac{1}{n}.$$

The mid 1970s brought with it the hand-held, microchip-centred, battery-powered, comparatively cheap calculator, thereby bringing to an end the role of logarithms and the slide rule as calculative aids. Yet the appearance of them in a piece of mathematics is seldom a cause for surprise. Anyone who has studied calculus would see them materialize time and again, quite probably in the expression for the integral of some function or in their role as the inverse of the exponential function, with e vying with π for constant supremacy. They can also arise without warning in situations that seem remote from their influence, and when they do so they exercise a surprising control in unexpected places—as we shall see: we will also see that the harmonic series, and others related to it, enjoy an important existence of their own.

The book naturally separates into two parts: Chapters 1–11 might be described as 'theory', and the remainder as 'practice'.

In the 'theory' part we are concerned with definitions and some consequences of them, methods to approximate, and to some extent, with preparation for the remaining chapters. We start by looking at the peculiar way in which logarithms were initially defined, a way which reveals the immense intellectual effort that must have been invested to turn multiplication into addition, to utilize an idea from the old world that helped to usher in the new. The harmonic series, with its three peculiar properties, is discussed and then its specializations and generalizations, before looking more closely at that definition of γ and having done that, and having convinced ourselves that the number actually exists, at ways of approximating its value, using both decimal and fractional methods. Among all of this we prove a barely credible result about co-prime integers and establish an identity (of Euler's) that holds the key to the modern study of prime numbers.

The later chapters, which are devoted to 'practice', look at some of the ways in which the three objects of our attention can appear in mathematics, and to some extent, in applications of it. Gamma's varied roles in analysis and number theory are mentioned, some surprising appearances of the harmonic series are discussed, and three such of logarithms. The finale is really just another application of logarithms, but since the application is the Prime Number Theorem, leading to the Riemann Hypothesis (neither of which we prove!), it is deservedly singled out. It is inevitable that our journey reaches mathematics that is 'worthy of serious consideration', as Euler himself said of γ, but none is more worthy than that celebrated Prime Number Theorem and that awesome Riemann Hypothesis; the first harnesses the wayward behaviour of the primes, the second adds finesse to that control by asking about the zeros of a function that seems to have none, but which stands alone as the greatest problem in mathematics today.

How difficult is the mathematics? That of course is a subjective matter. Certainly, we have not shied away from the use of symbols, since to do so would have condemned us merely to talking about mathematics rather than actually doing it. Yet, there are few really advanced techniques used, it is more that in some places simple ideas have been used in advanced ways. Mathematics makes a nice distinction between the usually synonymous terms 'elementary' and 'simple', with 'elementary' taken to mean that not very much mathematical knowledge is needed to read the work and 'simple' to mean that not very much mathematical ability is needed to understand it. In these terms we think the content is often elementary but in places not so very simple. The reader should expect to make use of a pen and paper in many places; mathematics is not a spectator sport! The approach is reasonably rigorous but informal, as this is no textbook, it is more a context book of mathematics in which the reader is asked to take time out from studying the mathematics to read a little around it and about the mathematicians who produced it or of the times in which they lived; sometimes in detail but other times just a few lines and then not always, as this is no history of mathematics book either; it merely acknowledges that mathematics comes from mathematicians, not books, and seeks to bring a sometimes shadowy figure forward to share the prominence of his ideas, and to give some sort of feel for the way in which those ideas developed over time.

The exception to the 'elementary' classification is some of the content of the final chapter on the Riemann Hypothesis; necessarily, this involves some complex function theory and in particular complex differentiation and integration. To those who have met these ideas the work should present few problems, but to those who have not they will look rather frightening; if so, simply ignore them or better still try to find out about them since they are a most glorious and powerful construction; a 'crash course' in some elements of complex function theory is included in Appendix D. The Riemann Hypothesis really is the greatest unsolved problem in mathematics, so it shouldn't be surprising that it is neither 'elementary' nor 'simple'; if the chapter entices hunger in some to get to grips with Cauchy's great invention it will have justified itself on that ground alone.

We hope that the material will appeal to a variety of people who have a little probability and statistics and a good calculus course behind them, and before that a rigorous course in algebra, if such a thing still exists: the motivated senior secondary student, who may well be seeing many of the ideas for the first time, the college student for whom the text may put flesh on what can sometimes be dry bones, the teacher for whom it might be a convenient synthesis of some nice ideas (and maybe the makings of a talk or two), and also those who may have left mathematics behind and who wish to remind themselves why they used to find it so fascinating. The reader will judge to what extent this book achieves its aim: to explain interesting mathematics interestingly.

The names of many mathematicians appear, names that should bring wonder to anyone interested in the subject and its history, but it is that name Euler that will force itself onto the page more than any other. It is not that we happen to pass through the mathematical territory to which he holds title, but more that it would be difficult, if not impossible, to go far in any mathematical direction without feeling his influence. For example, much of the notation that we now take for granted originates from him; in particular, $e, i, f(x), \sum, \Delta, \sin x, \cos x$, etc., as well as the standard manner of labelling a triangle, with the vertex the capital letter corresponding to the opposite side's small letter. It can be hard to appreciate, or easy to forget, just how many important ideas his name is associated with or perhaps even attached to; he invented many vastly important concepts and touched every known area of the subject—and everything he touched he adorned. According to R. Calinger, 'Euler's books and memoirs, of which 873 have so far been listed, comprise approximately a third of the entire corpus of research on mathematics and mechanics, both rational and engineering, published from 1726 to 1800.' The *Opera Omnia*, his collected works, has reached 74 volumes of 300 to 600 pages each; the final part has still to be finished and will comprise at least another seven volumes. Looking up 'Euler' in the index of a mathematics or history of mathematics book can be a frustrating experience, as the eye is routinely confronted with a block of page references, sometimes unspecified, at other times separated into a list, which might begin,

> Euler angles, Euler triangle, Euler characteristic, Euler's identity, Euler circle, Euler circuit, Euler–Mascheroni constant, Euler line, Euler numbers, Euler's first integral, Euler's second integral, Euler polynomials, Euler's Totient function, etc.,

and continue for dozens more entries.

And perhaps all that was needed was to know how to pronounce his name: 'Oiler'.

The noun 'genius' has been defined as 'exalted intellectual power, instinctive and extraordinarily imaginative and creative capacity'. Extravagant use of the word serves only to dilute its meaning or to bring into question the judgement of the author, but we have used it already and will risk employing it on a number of other occasions, no more fittingly than with Euler, safe in the conviction that if he was not a genius and these people were not geniuses then none have yet been born. Yet, to the majority, his name is probably as mysterious as his constant. He breathed life into γ through his Zeta functions (the generalizations of H_n), the summation of one of which was to become a long-standing problem—described as 'the despair of analysts'—until Euler's outrageous solution put an end to it.

With Euler and with those who preceded him and to some extent those who followed him we will deal with times remote from the modern years of 'publish or perish' and in consequence primacy over an initiative is often far from easy to establish; it might depend on a note to a contemporary or a recorded comment more often than an article in a learned journal, and even then that article might appear years after the actual breakthrough (the controversy surrounding the discovery of the calculus by Newton and Leibnitz stands as an infamous example of the problems that can arise). We hope that the reader will understand if the story is not always complete, and agree that where it is not complete it is at least representative.

Dr Urs Burckhardt, President of the Euler Commission, has written, 'Indeed, through his books, which are consistently characterized by the highest striving for clarity and simplicity and which represent the first actual textbooks in the modern sense, Euler became the premier teacher of Europe not only of his time but well into the 19th century.' Euler, as ever, provides a target too distant to reach, or even clearly to see, yet the pleasures (and frustrations) of achieving a fresh understanding of old ideas and realizations of new ones has proved marvellously invigorating and has brought with it the reminder that the best way of learning is by teaching, whether it be by the spoken or written word. We hope that the reader will share our enthusiasm as we take brief excursions though countries, centuries, lives and works, unfolding the stories of some remarkable mathematics from some remarkable mathematicians.

The Logarithmic Cradle

The use of this book is quite large, my dear friend,
No matter how modest it looks,
You study it carefully and find that it gives
As much as a thousand big books.

John Napier (1550–1617)

1.1 A MATHEMATICAL NIGHTMARE — AND AN AWAKENING

In an age when a 'computer' is taken to mean a machine rather than a person and calculations of fantastic complexity are routine and executed at lightning speed, constricting difficulties with ordinary arithmetic seem (and are) extremely remote. The technological freeing of mathematics from the manacles of calculation is very easy to take for granted, although the freedom has been newly won; as recently as the mid 1970s, a mechanical calculator, slide rule or table of logarithms would have been used to perform anything other than the most basic calculations—and the user would have been grateful for them. In the early 17th century none of these aids existed, although it was a period of massive scientific advance in many fields, progress that was increasingly and frustratingly hampered by the overwhelming difficulties of elementary arithmetic. Addition and subtraction were quite manageable, but how could the much more difficult tasks of multiplication and division be simplified, let alone the important but formidably challenging processes of root extraction?

Ancient civilizations had tackled the problem. For example, the Babylonians were known to have used the equivalent of $ab = \frac{1}{4}((a + b)^2 - (a - b)^2)$, which, with a table of squares, provides some calculative help. The 16th century brought with it more sophisticated ideas, particularly one using the unlikely device of trigonometric identities, the brainchild of two Dutch mathematicians named Wittich and Clavius. Various relationships between the trigonometric definitions were appearing throughout Europe and, for example, François Vièta (1540–1603) is known to have derived (among others) the formula

$$\sin x \cos y = \frac{1}{2}(\sin(x + y) + \sin(x - y)), \tag{1.1}$$

1

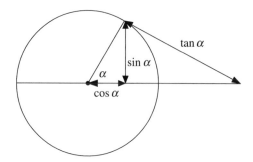

Figure 1.1. The medieval view of the trigonometric functions.

where the sine of an angle meant the length of the semi-chord of a circle, as in Figure 1.1; it therefore depended on the radius of the defining circle. In spite of the difficulties, extensive tables of the trigonometric functions were available, accurate to 12 or more decimal places (although written as integers by choosing a large whole number for the radius of the circle), their painstaking compilation motivated by practical problems in navigation, calendar construction and astronomy—and with the ingenuity of Wittich and Clavius they were set to other work.

The identity (1.1), with a set of trigonometric tables and scaling, could be used to convert multiplication to addition and subtraction (and division by 2); a technique known as 'prosthaphaeresis' (from the Greek for addition and subtraction). Division could be managed in much the same way, using identities for secants and cosecants. This slender aid found use wherever it became known and nowhere more effectively than in the astronomical observatories of Europe, none more prestigious than Uraneinborg (Castle in the Sky), on the island of Hven, where the Swedish–Danish Astronomer Royal, Tycho Brahe (1546–1601), lived and worked. And here appears a romantic story, bringing about a delicious serendipity. In 1590, James VI of Scotland (later to become James I of England) sailed to Denmark to meet his prospective wife (Anne of Denmark) and was accompanied by his physician, a Dr John Craig. Appalling weather conditions had forced the party to land on Hven, near to Brahe's observatory, and quite naturally the great astronomer entertained the distinguished party until the weather cleared, partly by demonstrating to them the process of prosthaphaeresis. Dr Craig was Scottish and he had a particular friend who lived near Edinburgh: one John Napier.

John Napier, Baron Merchiston, believed that the world would end between 1688 and 1700, and published his belief in a 1593 polemic on Catholicism entitled *A Plaine Discovery of the Whole Revelation of St. John*; its main thesis was that the Pope was the Antichrist. Since the book ran to 21 editions (10 in his own lifetime), he had some justification in believing that this would be his greatest claim to posterity (such as there was to be of it); of course, he was wrong (on both counts) and it is for his Table (or Canon) of Logarithms and the two explanations of them (the 1614 *Descriptio* and the 1619 posthumously published *Constructio*) that he is best remembered. A massively committed Protestant, but no 'crank', he found time from his contributions to the religious and political ferment of the day to efficiently manage his considerable estates, present prophetic (and surprisingly accurate) ideas for machines of war (what we would call the machine gun, the tank and the submarine) and, of course, to study mathematics. The private manuscript 'De Arte Logistica' (which was not published until 1839) provides an insight into his mathematical interests, which included a study of equations (and even consideration of imaginary numbers) and general methods for the extraction of nth roots.

The relationship between arithmetic and geometric behaviour, which we would now write as $a^n \times a^m = a^{n+m}$, had been understood since antiquity; it is seen (for m and n positive integers) on Babylonian tablets and also in *The Sandreckoner* of the great Archimedes of Syracuse (278–212 B.C.), which we will mention again later on p. 93. In this treatise, which was dedicated to his relative, King Gelon of Syracuse, he constructed a systematic method for representing arbitrarily large numbers, using the number of grains of sand in the known universe as a tangibly large number; the work provides the first hint of the nature of logarithms. In its own way the identity also converts multiplication to addition; now, through his friend, Napier knew that with ingenuity more calculative aid was possible and, setting aside his study of arithmetic and algebra, he sought to improve the lot of scientists of his day, and in effect using this property of exponents. Twenty years later he had succeeded. In his own words, from the preface to the *Descriptio*:

> Seeing there is nothing (right well-beloved Students in the Mathematics) that is so troublesome to Mathematicall practise, nor that doth more molest and hinder Calculators, than the Multiplications, Divisions, square and cubical Extractions of great numbers, which besides the tedious expense of time are for the most parte subject to many slippery errors. I began therefore to consider in my minde by what certaine and ready Art I might remove those hindrances. And having thought upon many things to this purpose, I found at length some excellent briefe rules to be treated of (perhaps) hereafter. But amongst all, none more profitable than this which together with the hard and tedious Multiplications, Divisions, and Extractions of rootes, doth also cast away from the worke it selfe, even the very numbers themselves that are to be multiplied, divided and resolved into rootes, and putteth other numbers in their place which performe as much as they can do, onely by Addition and Subtraction, Division by two or Division by three...

3

1.2 THE BARON'S WONDERFUL CANON

We will tread some of Napier's path by annotating a small part of the second publication, *The Construction of the Wonderful Canon of Logarithms*, usually abbreviated to the *Constructio*.

It begins with 60 numbered paragraphs that combine to explain his approach, provide a limited table of logarithms, and give instruction on how to make more extensive ones.

> (1) A Logarithmic Table is a small table by the use of which we can obtain a knowledge of all geometrical dimensions and motions in space, by a very easy calculation.

His first sentence suggests Napier's interest in the practical applications of logarithms, and perhaps most particularly their usefulness to astronomers such as Tycho Brahe, with whom he corresponded during their development. Although his invented word 'logarithm' (a compound from the Latin words meaning ratio and number) appeared in the title, he used 'artificial number' in the body of the text and 'Logarithmic Table' was written 'Tabula Artificialis'.

> It is deservedly called very small, because it does not exceed in size a table of sines; very easy, because by it all multiplications, divisions, and the more difficult extractions of roots are avoided; for by only a very few most easy additions, subtractions and divisions by two, it measures quite generally all figures and motions.

The 'modesty' of the volume is referred to and he makes clear the arithmetic advantages of using logarithms, but refers only to square roots here.

> It is picked out of numbers progressing in continuous proportion.

A hint as to the method.

> (2) Of continuous progressions, an arithmetical is one which proceeds by equal intervals; a geometrical, one which advances by unequal and proportionally increasing or decreasing intervals...

His definition of arithmetic and geometric progressions, after which he lists several examples.

> (3) In these progressions we require accuracy and ease in working. Accuracy is obtained by taking large numbers for a basis; but large numbers are most easily made from small by adding ciphers. Thus, instead of 100 000, which the less experienced make the greatest sine, the more learned put 10 000 000, whereby the difference of all sines is better expressed. Wherefore also we use the same for radius and for the greatest of our geometrical proportionals.

A cipher is a zero and attaching them to the right-hand side of a number does indeed increase the size of the number. To avoid the use of fractions, it was customary to 'change the units' (rather like using millimetres rather than metres). The 'greatest sine' is the radius of the circle, achieved when $\alpha = 90°$ in Figure 1.1, and Napier chooses to represent this as 10^7 units, rather than a mere 10^5.

(4) In computing tables, these large numbers may again be made still larger by placing a period after the number and adding ciphers. Thus in commencing to compute 10 000 000 we put 10 000 000.000 000 0, lest the most minute error should become very large by frequent multiplication.

Here he acknowledges the dangers of compounding rounding errors and introduces the use of the decimal point to help cope with them. His idea is that, even though the final logarithm will be rounded off to an integer, the intermediate calculations should involve as much accuracy as possible.

Extending the Hindu place-value number system to include decimals had been one of the conceptual and notational difficulties of mathematics, and one of its most important developments. In 1530 one Christof Rudolff (1499–1545) used a form of decimal fractions in a published collection of arithmetic examples; he also brought to the mathematical world the radical sign $\sqrt{}$ for the square root. It was, though, the multi-faceted Dutch scientist, Simon Stevin (1548–1620), who is accepted to have championed the use of decimal places more than anyone before him, since in 1585 he produced the first known systematic presentation of the rules for manipulating them in the treatise *De Thiende*. His ideas soon reached a far greater audience when the book was quickly translated from Dutch to French to become *La Disme*, which has the subtitle 'Teaching how all computations that are met in business may be performed by integers alone without the aid of fractions'. More of a pamphlet than a book, there is a resonance with Napier's thoughts in the quite splendid introduction.

> To astrologers, surveyors, measurers of tapestry, gaugers, stereometers in general, mintmasters and to all merchants, Simon Stevin sends greeting.

> A person who contrasts the small size of this book with your greatness, my most honourable sirs to whom it is dedicated, will think my idea absurd, especially if he imagines that the size of this volume bears the same ratio to human ignorance that its usefulness has to men of your outstanding ability; but in so doing he will have compared the extreme terms of the proportion which may not be done. Let him rather compare the third term with the fourth.

> What is it here that is being propounded? Some wonderful invention? Hardly that, but a thing so simple that it scarce deserves the name invention; for it is as if some stupid country lout chanced

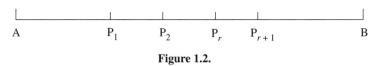

Figure 1.2.

upon great treasure without using any skill in the finding. If any one thinks that, in expounding the usefulness of decimal numbers, I am boasting of my cleverness in devising them, he shows without doubt that he has neither the judgement nor the intelligence to distinguish simple things from difficult, or else that he is jealous of a thing that is for the common good. However this may be, I shall not fail to mention the usefulness of these numbers, even in the face of this man's empty calumny. But, just as the mariner who has found by chance an unknown isle, may declare all its riches to the king, as, for instance, its having beautiful fruits, pleasant plains, precious minerals etc., without its being imputed to him as deceit; so may I speak freely of the great usefulness of this invention, a usefulness greater than I think any of you anticipates, without constantly priding myself on my achievements.

His notation varied from very to reasonably cumbersome, for example, $3 \odot 1 \odot 4 \odot 2 \odot$, $3/\underline{142}$ and $3^{\underline{142}}$. Napier was not consistent with his own notation but his use of the decimal point in the *Constructio* was to bring about a standardization, at least to some extent; even today, the Americans would usually write 3.142, the Europeans 3,142 and the English 3·142. Certainly, decimal is far superior to fractional notation when comparing sizes—and composing tables—and it was Napier's tables of logarithms that did most to popularize this crucial initiative.

(5) In numbers distinguished thus by a period in their midst, whatever is written after the period is a fraction, the denominator of which is unity with as many ciphers after it as there figures after the period.

Thus $10\,000\,000.04$ is the same as $10\,000\,000\frac{4}{100}\ldots$

The original *Descriptio* did not include explicit use of decimals. He continues to give several examples of the meaning of decimal notation.

The next paragraph to interest us is

(25) Whence a geometrically moving point approaching a fixed one has its velocities proportionate to its distances from the fixed one...

A lengthy rhetoric follows, referring to the equivalent of Figure 1.2, to establish that if a point P starts at A and moves continuously towards B in such a way that $BP_r : BP_{r+1}$ is constant (and therefore moving 'geometrically'), then that constant is the ratio of the point's velocities at P_r and P_{r+1}: that is, $V_r : V_{r+1} = BP_r : BP_{r+1}$.

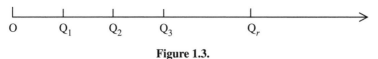

Figure 1.3.

To establish this, Napier considered the motion of P over equal time intervals of length t and implicitly approximated the varying speed over each interval by its value at its starting point, as we might do in step-by-step solutions of differential equations. In modern notation, suppose that at some stage P is at position P_r and that at some fixed time t later it is at P_{r+1}, then $BP_r = BP_{r+1} + P_r P_{r+1} = BP_{r+1} + V_r t$, using the above approximation. Since $BP_{r+1}:BP_r = k$, $BP_r = kBP_r + V_r t$ and $V_r = (1/t)(1 - k)BP_r$. Of course, this means that $V_{r+1} = (1/t)(1 - k)BP_{r+1}$ and so $V_{r+1}:V_r = BP_{r+1}:BP_r$, as required. In a sense he was, of course, on subtle mathematical ground here, with the hint of instantaneous velocity, a concept that was to be dealt with by Newton seventy years in the future.

(26) The logarithm of a given sine is that number which has increased arithmetically with the same velocity throughout as that with which radius
began to decrease geometrically, and in the same time as radius has
decreased to the given sine.

This crucial paragraph defines his version of logarithm. Firstly, referring to Figure 1.2, AB is taken to be the 'radius' of length 10^7 and the possible values of $\sin \alpha$ are represented by distances along the line from B, with the whole 10^7 at A and 0 at B. The point P starts at A and moves towards B with a speed numerically equal to its distance from B, which means that its initial speed is 10^7 and its final speed 0 (although this is impossible to achieve). The key to the whole matter is his introduction of a second, infinite line to represent the motion of another point Q, starting at the same time as P from an origin O but moving continuously with a constant velocity of 10^7 (see Figure 1.3). He defines a set of points Q_r along this second line by the following: Q_r is the point reached by Q just as P reaches P_r; since the time intervals are equal and Q moves at constant speed, the intervals between the Q_r will all be equal and its motion 'arithmetic'. The OQ_r are defined to be the logarithm of the corresponding BP_r, which we will write as $OQ_r = \text{NapLog}(BP_r)$.

If we start to construct his table of logarithms, the implications of all this become more clear.

In the first time interval t, P moves to P_1, where $BP_1 = 10^7 - AP_1 = 10^7 - 10^7 t = 10^7(1 - t)$, approximating its speed over the interval by its initial speed of 10^7. During this time, Q will have moved to Q_1, where $OQ_1 = 10^7 t$, which means that $\text{NapLog}\{10^7(1 - t)\} = 10^7 t$. Repeating this analysis for the next time interval gives $BP_2 = 10^7 - AP_2 = 10^7 - (AP_1 + P_1 P_2) = 10^7 - 10^7 t - V_1 t = 10^7(1 - t) - V_1 t$. Now we use the result of the previous paragraph to get $V_1:10^7 = BP_1:10^7$ and therefore $V_1 = BP_1 = 10^7(1 - t)$, which means that $BP_2 = 10^7(1 - t) - 10^7(1 - t)t = 10^7(1 - t)^2$. Since $OQ_2 = 10^7 \times 2t = 2(10^7 t)$, we have that $\text{NapLog}\{10^7(1 - t)^2\} = 2(10^7 t)$. And so the process continues. In effect, he then takes $t = 1/10^7$ to get

$$\text{NapLog}\left\{10^7\left(1 - \frac{1}{10^7}\right)^1\right\} = \text{NapLog}(9\,999\,999) = 1,$$

$$\text{NapLog}\left\{10^7\left(1 - \frac{1}{10^7}\right)^2\right\} = \text{NapLog}(9\,999\,998) = 2,$$

and, in general,

$$\text{NapLog}\left\{10^7\left(1 - \frac{1}{10^7}\right)^r\right\} = r, \quad r \in \mathbb{N}.$$

And using the fact that the motion is continuous,

$$\text{NapLog}\left\{10^7\left(1 - \frac{1}{10^7}\right)^L\right\} = L \quad \text{for any positive } L.$$

The last paragraph that we will consider is

(27) Whence nothing is the logarithm of the radius...

$BA = 10^7$ is the 'radius' and with $P = A$, $Q = O$, $\text{NapLog}(10^7) = 0$.

The process can be thought of as taking powers of $(1 - 1/10^7)$, a number close to 1, which makes the powers close together and interpolation between them comparatively accurate; the factor of 10^7 eliminates the decimals. The *Constructio* continues to give methods of interpolation to fill in the gaps along AB, and in particular Napier notes that the geometric mean of two numbers corresponds to the arithmetical mean of their logarithms, which is true since if $L_1 = \text{NapLog}\, N_1$ and $L_2 = \text{NapLog}\, N_2$, then

$$N_1 = 10^7\left(1 - \frac{1}{10^7}\right)^{L_1},$$

$$N_2 = 10^7\left(1 - \frac{1}{10^7}\right)^{L_2},$$

$$\sqrt{N_1 \times N_2} = \sqrt{10^7\left(1 - \frac{1}{10^7}\right)^{L_1} \times 10^7\left(1 - \frac{1}{10^7}\right)^{L_2}}$$

$$= 10^7\left(1 - \frac{1}{10^7}\right)^{(L_1+L_2)/2}$$

and so

$$\text{NapLog}(\sqrt{N_1 \times N_2}) = \tfrac{1}{2}(L_1 + L_2).$$

It is easy to see that another important observation for their construction holds: if $N_1{:}N_2 = N_3{:}N_4$, then $\text{NapLog}(N_1) - \text{NapLog}(N_2) = \text{NapLog}(N_3) - \text{NapLog}(N_4)$.

A small variation of the reasoning exposes the use of logarithms as a calculative aid:

$$N_1 \times N_2 = 10^7\left(1 - \frac{1}{10^7}\right)^{L_1} \times 10^7\left(1 - \frac{1}{10^7}\right)^{L_2}$$
$$= 10^7 \times 10^7\left(1 - \frac{1}{10^7}\right)^{L_1+L_2},$$

which makes

$$\frac{N_1 \times N_2}{10^7} = 10^7\left(1 - \frac{1}{10^7}\right)^{L_1+L_2}$$

and so

$$\text{NapLog}\left(\frac{N_1 \times N_2}{10^7}\right) = L_1 + L_2 = \text{NapLog}\,N_1 + \text{NapLog}\,N_2,$$

and the familiar multiplicative law of logarithms emerges in a modified but still useful form, differing only in the position of the decimal point: multiplication had been transformed into addition. Napier noted that this 'functional relationship' satisfied by his logarithms allows the logarithm of any whole number to be calculated from knowing the logarithms of its prime factors, with primes making the first appearance of many in this book. As the gaps were filled, so the multiplication of a greater variety of numbers could be changed into their addition and the 'Wonderful Canon' be seen as the momentous aid to calculation that it was; Napier will forever be remembered as the discoverer of logarithms. He had built a new bridge that connected problems of multiplication and division to problems of addition and subtraction; 'prosthaphaeresis' had come of age.

Unfortunately, the name of the Swiss Jobst Bürgi (1552–1632) has slipped into obscurity, yet he had independently thought of the same idea, with a method differing only in detail. The most famous clockmaker of his time, a maker of scientific instruments and algebraic tutor to Johannes Kepler (1571–1630), he had published his method only in 1620, although it is clear that he was thinking of the ideas as early as 1588. It would take until 1707 for the birth of another Swiss who would leave an indelible mark on logarithms, and almost all other branches of mathematics: Euler.

The *Descriptio* opens with the verse at the head of this chapter, which plainly and amusingly demonstrates Napier's optimistic view of his invention, and indeed it met with immediate and considerable acclaim, convincingly summed up in the words of John Keill (1672–1721), Fellow of The Royal Society and Savilian Professor of Astronomy at Oxford:

The Mathematicks formerly received considerable Advantages; first by the Introduction of the Indian Characters, and afterwards by the Invention of Decimal Fractions; yet has it since reaped as least as much from the Invention of Logarithms, as from both the other two. The Use of these, every one knows, is of the greatest Extent, and runs through all Parts of Mathematicks. By their Means it is that Numbers almost infinite, and such as are otherwise impracticable, are managed with Ease and Expedition. By their assistance the Mariner steers his Vessel, the Geometrician investigates the Nature of higher Curves, the Astronomer determines the Places of the Stars, the Philosopher accounts for other Phenomena of Nature; and lastly, the Usurer computes the Interest of his Money.

The work found the particular affections of Henry Briggs (1561–1630), who had become the first Professor of Geometry at Gresham College, London, in 1596; in 1620 he was to become the first occupant of the Savilian Chair of Geometry in Oxford; later we will meet the great G. H. Hardy, who held that prestigious post some 300 years later—and offered it as a prize! Briggs's interest in the study of eclipses in particular and calculative aids in general naturally attracted him to Napier's idea and in a letter dated 10 March 1615 to his friend James Ussher, he wrote

... wholly employed about the noble invention of logarithms, then lately discovered... Napper, lord of Markinston, hath set my head and hands a work with his new and admirable logarithms. I hope to see him this summer, if it please God, for I never saw a book which pleased me better or made me more wonder.

The meeting did take place that summer, with Briggs the guest of Napier for a month, and another followed in 1616; Napier's death in April 1617 prevented the planned arrangement for a third year. Over that time they discussed variations on the idea and came to agree on the suggestion of Napier that '0 should be made the logarithm of 1 and 100 000 &c the logarithm of the radius'. The *Constructio* continues with an appendix by Briggs (it was he who undertook to arrange the publication of its London edition) entitled 'On the Construction of another and better kind of Logarithms, namely one in which the Logarithm of unity is 0'. That important step taken, he continued in the first paragraph with '... and 10 000 000 000 as the logarithm of either one tenth unity or ten times unity...'; the final form had yet to be reached. In the end, of course, it was to be that the logarithm of 1 would be 0, and the logarithm of 10 would be 1, and so the tables of logarithms that were to be used for the next 350 years, the *Briggsian logarithms*, came into being.

With Napier's decline and death, it fell to Briggs to calculate the new tables and as early as 1617 he had published *Logarithmorum Chilias Prima*, consisting of the logarithms of the natural numbers from 1 to 1000. In 1624 he published the formidably detailed *Arithmetica Logarithmica*, in which he developed far more comprehensive tables and formulated means of calculating whole classes of logarithms (and of putting them to use). Of course, gaps remained and the calculations involved in filling them could be prohibitive; Edward Wright, a translator of Napier's work, remarked that sometimes finding the logarithm of a number was more troublesome that performing the calculation without them! Briggs even suggested that the logarithms should be computed by teams of people, and he offered to supply specially designed paper for the purpose.

It is interesting to note that first recorded appearance of '×' for multiplication appeared in an anonymous appendix to Edward Wright's 1618 translation of the *Descriptio*, thought to have been authored by William Oughtred (1574–1660), the inventor of the slide rule.

1.3 A TOUCH OF KEPLER

One of the most immediate and significant uses to which logarithms were put was, unsurprisingly, in astronomy. In 1601, on the death of the fractious Brahe, Kepler was promoted to take his place. Not only did he inherit his master's prestigious position but also his voluminous and incredibly accurate data, which he used to help him conduct his 'war with Mars', a war that he eventually won and from which he extracted his first two laws of planetary motion.

1. Planets move in ellipses, with the Sun at one focus and the other empty.

2. The radius vector describes equal areas in equal times.

The results relating to Mars were published in *Astronomia Nova* of 1609 and were later extended to the other planets, but his suspicion that there was a simple law relating the size of the orbits to the period of the planets remained just that for many years. In his own words:

> ...and if you want the exact moment in time, it was conceived mentally on 8th March in this year one thousand six hundred and eighteen, but submitted to calculation in an unlucky way, and therefore rejected as false, and finally returning on the 15th of May and adopting a new line of attack, stormed the darkness of my mind. So strong was the support from the combination of my labour of seventeen years on the observations of Brahe and the present study, which conspired together, that at first I believed I was dreaming, and assuming my conclusion among my basic premises. But it is absolutely certain and exact that the proportion between the periodic times of any two planets is precisely the sesquialterate proportion of their mean distances...

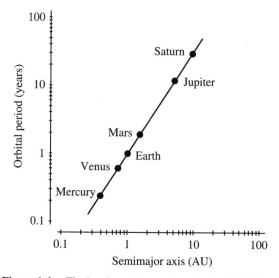

Figure 1.4. The log–log plot revealing Kepler's third law.

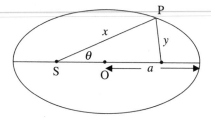

Figure 1.5. A planet's elliptical orbit.

He had published the result in his 1619 *Harmonice Mundi* as a late but important addition to the book, which was already at press when he finally discovered that $T \propto D^{3/2}$. Put another way, 'the square of the period is proportional to the cube of the average distance of the planet from the Sun'. How did he finally manage to discover the law? It is not clearly documented, but in 1616 he had read the *Descriptio*, and as a result logarithms would surely have helped him to see the hidden pattern.

In modern terms, a $\log T$–$\log D$ plot would yield the straight line in Figure 1.4: it is all so obvious, retrospectively!

The D is more easily realized as the length of the semimajor axis of the elliptical orbit. A little calculus establishes this.

Referring to Figure 1.5, by the definition of an ellipse, $x + y = 2a$, and so

$$\int_0^{2\pi} x + y \, d\theta = \int_0^{2\pi} 2a \, d\theta = 4\pi a.$$

So,

$$\int_0^{2\pi} x \, d\theta + \int_0^{2\pi} y \, d\theta = 2 \int_0^{2\pi} x \, d\theta = 4\pi a.$$

Therefore, the average value of the distance of the planet from the Sun is

$$\frac{1}{2\pi} \int_0^{2\pi} x \, d\theta = \frac{2\pi a}{2\pi} = a.$$

Whatever the facts with the third law, it is certain that Kepler used (and indeed justified and developed) logarithms for the production of the 1628 *Rudolphine Tables* of planetary positions, which itself contained a set of his own form of logarithms to eight-figure accuracy. In the words of Pierre Laplace (1749–1827) logarithms '. . . by shortening the labours, doubled the life of the astronomer'. A poetically formed, inaccurate, but powerfully revealing, observation.

1.4 A TOUCH OF EULER

The modern eye might well judge Napier's approach to logarithms as peculiar. They are defined in terms of the motions of points, there is no base and the logarithm of 10 000 000 was originally 0. All together, they seem so distant from what we now think of them to be, particularly as they are no longer used for the purpose for which they were invented: to calculate. Yet, there was an early suggestion of what we would consider logarithmic behaviour. Before 1636, Pierre Fermat (1601–1665) (among others, and whom we will meet again in a later chapter) had shown what we would write as

$$\int_0^a x^n \, dx = \frac{a^{n+1}}{n+1}$$

for all rational numbers $n \neq -1$, but the expression for the area under the rectangular hyperbola $y = 1/x$ continued to prove elusive. The first inkling of the connection with logarithms seemed to appear in 1647, in *Opus Geometricum. . .*, written by the Jesuit priest, Gregory St Vincent (1584–1667). The method of approximating areas by rectangles having an equal base was in common currency but here St Vincent used rectangles of equal area, adjusting their base accordingly.

Referring to Figure 1.6, since the areas of the first two rectangles are equal, $y_0(x_1 - x_0) = y_1(x_2 - x_1)$ and so

$$\frac{1}{x_0}(x_1 - x_0) = \frac{1}{x_1}(x_2 - x_1) : \frac{x_1}{x_0} - 1 = \frac{x_2}{x_1} - 1 \quad \text{and} \quad \frac{x_1}{x_0} = \frac{x_2}{x_1}.$$

This means that for the areas to increase arithmetically, the x-coordinates increase geometrically, with the strong suggestion of a logarithmic law connecting the area under $y = 1/x$ with x. In his *Waste Book* of 1664, the great

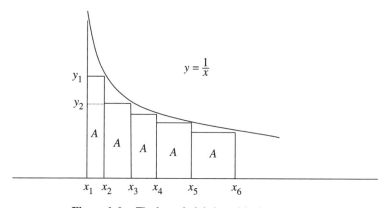

Figure 1.6. The hyperbola's logarithmic behaviour.

Isaac Newton (1642–1727) wrote, 'In ye Hyperbola ye area of it bears ye same respect as its Asymptote which a logarithme doth its number.' Both Newton and Nicholas Mercator (1620–1687) independently developed the idea, expanding $1/(1+x)$ as $1 - x + x^2 - x^3 + \cdots$ and integrating term by term to finish with the now standard expression $\log(1+x) = x - \frac{1}{2}x^2 + \frac{1}{3}x^3 - \cdots$ and thereby a much more convenient means of calculating logarithms. Logarithms were at once the 'artificial' numbers of Napier, the area under the rectangular hyperbola and the sum of an infinite series.

As we pass over many other individual contributions, the chasm separating the past and the present was filled by Euler more than anyone else; it was he who saw furthest. The synthesis of the several approaches to logarithms lay in Euler's definition of them, which appeared in his bestselling textbook on algebra *Complete Introduction to Algebra* of 1770:

220 Resuming the equation $a^b = c$, we shall begin by remarking that, in the doctrine of Logarithms, we assume for the root a, a certain number taken at pleasure, and suppose this root to preserve invariably its assumed value. This being laid down, we take the exponent b such that the power a^b becomes equal to a given number c; in which case this exponent b is said to be the logarithm of the number c...

221 We see, then, that the value of the root a being once established, the logarithm of any number, c, is nothing more than the exponent of that power of a, which is equal to c; so that c being $= a^b$, b is the logarithm of the power a^b.

His wording is cumbersome by modern standards but here is the definition of logarithm that confronts most people when they are introduced to them today. This remarkable book is made more remarkable still with the realization that at the time of its writing Euler was virtually blind; he dictated the manuscript to a servant, who was to function as his mathematical secretary. Later, he was to establish the idea of a function and one that approaches its modern definition, with $y = a^x$ a special case, and its inverse function defined as the logarithmic function. Earlier, in an article of 1749 with a title that transcends language barriers 'De la controverse entre Messers. Leibnitz et Bernouilli sur les logarithms négatifs et imaginaires', he used the series expansion for the natural logarithm and developed ideas of complex numbers to argue that the logarithm of any number is multivalued. Below is his jaw-dropping argument, which uses his famous logarithmic limit (independently discovered by Edmond Halley (1656–1742), of 'comet' fame). Using his terminology, w is an 'infinitely small' number and n an 'infinitely large' one, with l representing the logarithm.

Since w is 'infinitely small', $l(1 + w) = w$ and therefore $y = l(1 + w)^n = nw$. Now let $x = (1 + w)^n$, then $1 + w = x^{1/n}$ and $w = x^{1/n} - 1$, which means that $lx = y = n(x^{1/n} - 1)$. He then argued that there are n (complex) values of $x^{1/n}$ for any x and since n is an infinite number, there must be an infinite number of values of lx. He continued by pointing out that all but one of the values would involve $\sqrt{-1}$, presaging one of the most subtle ideas of the next century's complex function theory, the Riemann surface. This limit, $\ln x = \lim_{n \to \infty} n(x^{1/n} - 1)$, and the equally famed $e^x = \lim_{n \to \infty}(1 + x/n)^n$ both appear in his two-volume classic *Introductio in Analysin Infinitoram* of 1784, and in putting $x = -1$ in the second expression to get

$$\frac{1}{e} = \lim_{n \to \infty} \left(1 - \frac{1}{n}\right)^n,$$

we can begin to unravel Napier's thoughts.

Since $\text{NapLog}\{10^7(1 - 1/10^7)^L\} = L$, $\text{NapLog}\{10^7(1 - 1/10^7)^{10^7}\} = 10^7$. Now, 10^7 may not be 'infinity' but it is quite big enough for $(1 - 1/10^7)^{10^7}$ to be very accurately approximated by $1/e$ to get

$$10^7 = \text{NapLog}\left\{10^7\left(1 - \frac{1}{10^7}\right)^{10^7}\right\} \approx \text{NapLog}\left(10^7 \frac{1}{e}\right).$$

Now, if we scale down by a factor of 10^7, we have that $\text{NapLog}(1/e) \approx 1$, which suggests that $\text{NapLog}\, x$ might well be $\log_{1/e} x$.

With the use of the calculus, we can be precise.

In Figures 1.2 and 1.3, if we write $\text{PB} = x$, $\text{OQ} = y$ and the constant of proportionality 1 we have $dx/dt = -x$ and $dy/dt = 10^7$. The initial conditions are that when $t = 0$, $x = 10^7$ and $y = 0$. These give

$$\frac{dy}{dx} = \frac{dy}{dt}\frac{dt}{dx} = -\frac{10^7}{x}$$

15

and so $y = -10^7 \ln x + c$, where $0 = -10^7 \ln 10^7 + c$, which makes

$$y = -10^7 \ln x + 10^7 \ln 10^7 = 10^7 \ln \frac{10^7}{x} \quad \text{or} \quad \frac{y}{10^7} = \ln \frac{10^7}{x}.$$

If we notice that $\ln \lambda = \log_{1/e} 1/\lambda$, we finally have that

$$\frac{y}{10^7} = \log_{1/e} \frac{x}{10^7}.$$

So, Napier's logarithms really are a scaled-down version of logs, to the base $1/e$.

1.5 NAPIER'S OTHER IDEAS

Napier's major legacy is, then, a method of calculation that in its various forms has helped scientists and mathematicians over centuries to pursue their investigations and theories, relatively free of the tedium of arithmetic: logarithms modern role is deeper still, as we shall see. He bequeathed some other inheritances too.

Some of the most important practical geometrical problems of the time were involved with celestial navigation (with the Global Positioning System not even within the realms of science fiction) and therefore involved spherical triangles, with the Earth by then being acceptably round. Napier was recognized for two ideas connected with spherical trigonometry: a set of four identities useful for solving 'oblique' spherical triangles, given the name 'Napier's analogies', and two ingenious rules for remembering the ten formulae used in solving right-angled spherical triangles. Both are in use today and we list them below. They use the now standard labelling for a triangle (spherical or plane) that capital letters represent the vertices and the corresponding small letters the side opposite (as we mentioned in the introduction, yet another inheritance from Euler) and it should be borne in mind that any side of a spherical triangle can be thought of as the angle it subtends at the centre of the defining circle. In this notation, Napier's analogies are

$$\frac{\sin \frac{1}{2}(A - B)}{\sin \frac{1}{2}(A + B)} = \frac{\tan \frac{1}{2}(a - b)}{\tan \frac{1}{2}c}, \qquad \frac{\cos \frac{1}{2}(A - B)}{\cos \frac{1}{2}(A + B)} = \frac{\tan \frac{1}{2}(a + b)}{\tan \frac{1}{2}c},$$

$$\frac{\sin \frac{1}{2}(a - b)}{\sin \frac{1}{2}(a + b)} = \frac{\tan \frac{1}{2}(A - B)}{\cot \frac{1}{2}c}, \qquad \frac{\cos \frac{1}{2}(a - b)}{\cos \frac{1}{2}(a + b)} = \frac{\tan \frac{1}{2}(A + B)}{\cot \frac{1}{2}c}.$$

If the triangle is right-angled at A, the remaining five letters (two angles and three sides) can be arranged in order as points on a circle, as shown, with each point having two 'adjacent' points and two 'opposite points', which gives rise to Figure 1.7.

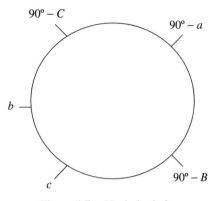

Figure 1.7. Napier's circle.

Napier's rules are, then,

- the sine of any point equals the product of the tangents of the adjacent points,

- the sine of any point equals the product of the cosines of the opposite points.

Moving around the circle gives five lots of two formulae.

Most famous of all is his other calculating device: *Napier's bones* (or *rods*). From them came the slide-rules of Oughtred, Gunter and Mannheim and, without the silicon chip, we would be using something based on them today. It seems certain that the idea stems from an ancient Arabic scheme for organizing, and therefore simplifying multiplication; the elegant *gelosia* or *grating* method. This (literally) romantic name is an allusion to the grid used in the method, which resembles a type of window lattice (or *gelosia*) through which a jealous spouse might peer unseen. The process starts with a blank design into which the two numbers to be multiplied are introduced; the example shows the product

$$3284 \times 6751 = 22\,170\,284.$$

The two numbers are written in the top and right semicircles (in bold in Figure 1.8), the individual products of the digits are then written in the diagonally split squares, forming a restricted 'times table'; the answer appears in the remaining semicircles (the underlined digits in the figure), having been formed by adding the digits diagonally, starting at the bottom right—carrying over where necessary.

His development of this idea was published in his *Rabdologia* of 1617 (from the Greek for 'rod' and 'collection'), the year of his death, and they became extremely popular; perhaps the rods served the needs of those for whom logarithms were too abstract an idea. He devised several variations; some capable

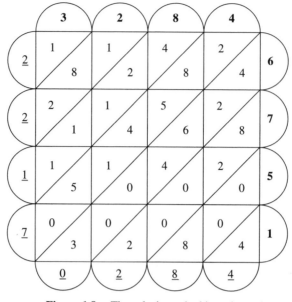

Figure 1.8. The gelosia method in action.

of root extraction, but it is the type that deals with basic multiplication and division that has been most widely remembered. With these, each of the 10 possible arithmetic rods comprises a digit at the top and the multiplication table for that digit permanently written below, in the same way as the *gelosia's* were written out each time; an 11th rod is simply the digits 1 to 9, written in order, as illustrated on the left in Figure 1.9. To multiply two numbers, represent one of them as a row of rods with the number forming the top row (of course, that means repeated digits require repeated rods); 5978 on the right in Figure 1.9. The index rod is then placed next to the set-up and the second number multiplied one digit at a time and the results added, taking account of decimal positions. For example, in the illustration, the digit 5 is being multiplied to give 29 890 using the same diagonal adding as with the *gelosia*.

In 1890, the French Civil Engineer, Henri Genaille, produced an elegant refinement which has become known as Genaille's rods and which removed the need to remember the carry; they quite literally allow the user to read off the answer to a product of a number with a single digit with no calculation whatever.

Finally, the *Rabdologia* contained yet another scheme for easing calculation: *Napier's abacus*. The use of chequered (chess) boards for calculating was well established by Napier's time and in his Abacus he used such a board with counters which take the move of a bishop or rook to perform all four arithmetic operations, as well as root extraction. For this he needed the ancient idea of

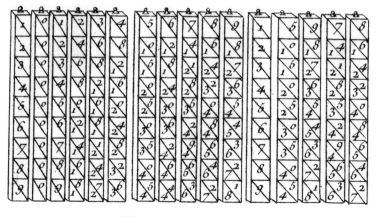

Figure 1.9. Napier's bones.

multiplying by doubling; in other words, writing numbers as powers of two. He would never have realized it, but in doing so he was using binary arithmetic, presaging the modern computer by some 350 years.

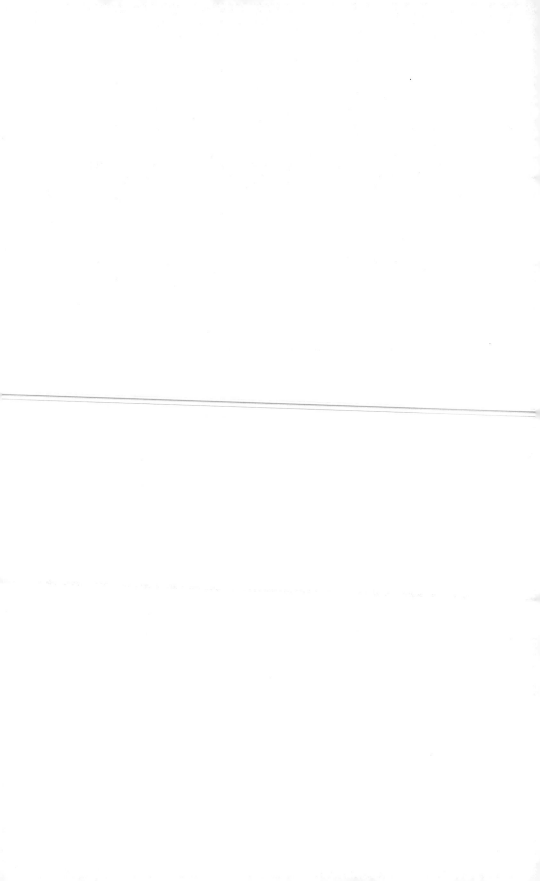

The Harmonic Series

Mathematicians are like lovers. Grant a mathematician the least principle, and he will draw from it a consequence which you must also grant him, and from this consequence another.

Bernard Le Bovier Fontenelle (1657–1757)

2.1 THE PRINCIPLE

On 11 July 1382, in the beautiful Norman city of Lisieux, Nicholas Oresme died at the age of 59; he had been the city's bishop since 1377. Born into the Late Middle Ages (in Allemagne in 1323), his scholarship extended from the development of the French language to taxation theory and his distinguished career included the Deanship of Rouen and being chaplain to King Charles V of France, for whom he translated Aristotle's *Ethics, Politics and Economics*. He taught the heliocentric theory of Copernicus over 100 years before Copernicus was born and suggested graphing equations nearly 200 years before the birth of Descartes; his treatise *De Moneta* brought him the soubriquet of the greatest medieval economist, but it is in his research in mathematics (it is probable that he was the first to use '+' for addition and it was he who, in his *Algorismus Proportionum*, extended index notation to fractional and negative powers) and in particular in infinite series that our interest lies. To be exact, we are concerned with his work on the harmonic series and his proof of a single property of it and in that specialization we consciously ignore almost everything that this great man achieved; it is rather like remembering the inimitable Carl Frederick Gauss (1777–1855) for a measurement of magnetic flux. The greatest mathematician of them all will reappear time and again throughout the next pages, but now it is Oresme's turn, but before that we establish a....

2.2 GENERATING FUNCTION FOR H_n

The definition of the harmonic series,

$$H_n = \sum_{r=1}^{n} \frac{1}{r} = 1 + \frac{1}{2} + \frac{1}{3} + \cdots + \frac{1}{n},$$

is equivalent to

$$H_r = H_{r-1} + \frac{1}{r}, \quad r > 1 \text{ and } H_1 = 1,$$

which can be used to establish the generating function:

$$\frac{1}{1-x} \ln\left(\frac{1}{1-x}\right) = \sum_{r=1}^{\infty} H_r x^r.$$

If we make no assumption about the H_r, we can multiply across by the $(1 - x)$ to get

$$-\ln(1 - x) = (1 - x) \sum_{r=1}^{\infty} H_r x^r$$

and then use Newton and Mercator's expansion of $\ln(1 - x)$ to get

$$\therefore x + \tfrac{1}{2}x^2 + \tfrac{1}{3}x^3 + \cdots = \sum_{r=1}^{\infty} H_r x^r - \sum_{r=1}^{\infty} H_r x^{r+1}, \quad \text{assuming } |x| < 1.$$

Comparing coefficients of x^r, we have

$$\frac{1}{r} = H_r - H_{r-1} \quad \text{for } r > 1,$$

$$\therefore H_r = H_{r-1} + \frac{1}{r} \quad \text{and } H_1 = 1,$$

the definition is recovered and the result is established.

In a letter dated 15 February 1671, James Gregory (1638–1675) wrote, 'As to yours, dated 24 Dec., I can hardly beleev, till I see it, that there is any general, compendious & geometrical method for adding an harmonical progression. . . '. To this day, we share Gregory's disappointment, as a formula for H_n for general n does not exist, nice though it would be to have. The simplicity of the definition of the series belies its subtlety and many consequences can be drawn from it. Below we give three.

2.3 THREE SURPRISING RESULTS

2.3.1 *Divergence*

No property is more unexpected than H_n's divergence, and it is this that Oresme proved; that is, as $n \to \infty$, $H_n \to \infty$, but so very slowly. The first 100 terms sum to $5.187\ldots$, the first 1000 to $7.486\ldots$ and the first $1\,000\,000$ to $14.392\ldots$; it is hard to believe that, for large enough n, H_n will exceed any chosen number, but such is the case; it would take a sensitive eye indeed to spot the divergence numerically. In 1968 John W. Wrench Jr calculated the exact minimum number of terms needed for the series to sum past 100; that number is 15 092 688 622 113 788 323 693 563 264 538 101 449 859 497. Certainly, he did not add up the terms. Imagine a computer doing so and suppose that it takes it 10^{-9} seconds to add each new term to the sum and that we set it adding and let it continue doing so indefinitely. The job will have been completed in not less than 3.5×10^{17} (American) billion years.

Oresme's celebrated proof, in modern notation, is shown below:

$$H_\infty = 1 + \frac{1}{2} + \left(\frac{1}{3} + \frac{1}{4}\right) + \left(\frac{1}{5} + \frac{1}{6} + \frac{1}{7} + \frac{1}{8}\right)$$

$$+ \left(\frac{1}{9} + \frac{1}{10} + \frac{1}{11} + \frac{1}{12} + \frac{1}{13} + \frac{1}{14} + \frac{1}{15} + \frac{1}{16}\right) + \cdots$$

$$> 1 + \frac{1}{2} + \left(\frac{1}{4} + \frac{1}{4}\right) + \left(\frac{1}{8} + \frac{1}{8} + \frac{1}{8} + \frac{1}{8}\right)$$

$$+ \left(\frac{1}{16} + \frac{1}{16} + \frac{1}{16} + \frac{1}{16} + \frac{1}{16} + \frac{1}{16} + \frac{1}{16} + \frac{1}{16}\right) + \cdots$$

$$= 1 + \frac{1}{2} + \frac{2}{4} + \frac{4}{8} + \frac{8}{16} + \cdots = 1 + \frac{1}{2} + \frac{1}{2} + \frac{1}{2} + \frac{1}{2} + \cdots,$$

which is, of course, divergent.

Inevitably, such a result has many proofs and we will consider two more. With the pursuit of elegance as motive:

$$H_\infty = 1 + \frac{1}{2} + \frac{1}{3} + \frac{1}{4} + \cdots = \frac{2}{2} + \frac{2}{4} + \frac{2}{6} + \frac{2}{8} + \cdots$$

$$= \left(\frac{1}{2} + \frac{1}{2}\right) + \left(\frac{1}{4} + \frac{1}{4}\right) + \left(\frac{1}{6} + \frac{1}{6}\right) + \left(\frac{1}{8} + \frac{1}{8}\right) \cdots$$

$$< \left(1 + \frac{1}{2}\right) + \left(\frac{1}{3} + \frac{1}{4}\right) + \left(\frac{1}{5} + \frac{1}{6}\right) + \left(\frac{1}{7} + \frac{1}{8}\right) + \cdots$$

$$= H_\infty, \quad \text{a contradiction.}$$

And with deference to Euler, whose part in this story (and so many others) is so great:

$$\int_{-\infty}^{0} \frac{e^x}{1 - e^x}\, dx = \int_{-\infty}^{0} e^x (1 - e^x)^{-1}\, dx$$

$$= \int_{-\infty}^{0} e^x (1 + e^x + e^{2x} + e^{3x} + \cdots)\, dx = \int_{-\infty}^{0} e^x + e^{2x} + e^{3x} + \cdots \, dx$$

$$= [e^x + \tfrac{1}{2}e^{2x} + \tfrac{1}{3}e^{3x} + \cdots]_{-\infty}^{0} = 1 + \tfrac{1}{2} + \tfrac{1}{3} + \cdots = [-\ln(1 - e^x)]_{-\infty}^{0},$$

which is clearly infinite when evaluated at the upper limit. There we have it, with its improper integral and non-legitimate binomial expansion; it can be tidied of course, but forcing the detail would blur its sweeping stylishness.

2.3.2 H_n is non-integral

The second surprise is that, even though H_n increases without bound, it manages to avoid all integers in doing so (apart from $n = 1$) and, more than this, any consecutive subseries of H_n is never an integer. That is, for positive integers m, n with $m < n$,

$$S_{mn} = \frac{1}{m} + \frac{1}{m+1} + \frac{1}{m+2} + \cdots + \frac{1}{n}$$

is never an integer.

The argument we give is delicate and a bit wordy, but the method of proof is to show that S_{mn} is a fraction with an odd numerator and an even denominator, which ensures that it cannot be integral. To this end, we need an intermediate result, which itself is a little surprising at first.

In any finite, consecutive subsequence of the sequence $1, 2, 3, \ldots$, there is a unique term with a highest factor of 2. That is, if we factorize each term and focus on the factors that are the powers of 2, there is only one with the term with the highest power of 2. The following argument establishes this.

If the sequence contains powers of 2, the term with the highest power of 2 is the number we seek. Otherwise, the sequence is contained strictly between two consecutive powers of 2, say 2^α and $2^{\alpha+1}$, that is $2.2^{\alpha-1}$ and $4.2^{\alpha-1}$, the highest power of 2 between them being $3.2^{\alpha-1}$; if the sequence contains this number, it is the number we seek, otherwise the sequence lies entirely within one of the two intervals, say $2.2^{\alpha-1}$ and $3.2^{\alpha-1}$, that is, $4.2^{\alpha-2}$ and $6.2^{\alpha-2}$, the highest power of 2 between them being $5.2^{\alpha-2}$. The process continues until the sequence contains one of the key numbers or is of length 2, in which case we select the even number.

Now suppose that we factorize each of the denominators of S_{mn} into a product of prime factors and select the unique term whose denominator contains the highest power of 2; call it $1/k$. When each term of S_{mn} is written as a fraction with denominator the least common multiple of all of these, $1/k$ must have an odd numerator and the numerators of all of the others must be even, consequently the numerator of S_{mn}, considered as a single fraction, must be odd; clearly the denominator is even and we have the result. Of course, taking $m = 1$ proves the result for H_n.

2.3.3 H_n is almost always a non-terminating decimal

We have that $H_1 = 1$, $H_2 = 1.5$ and $H_6 = 2.45$ and of course, since H_n is always a fraction, its decimal expansion must either be finite, as with these examples, or infinitely recurring. The final surprise is that, apart from these three cases, all of the other H_n are the infinitely recurring variety. Our proof of this remarkable fact will take us from comparatively shallow to very deep water, with the need of a most profound and significant result of number theory: the Bertrand Conjecture. In 1845 the French mathematician Joseph Bertrand (1822–1900) conjectured that for every positive integer $n > 1$, there exists at least one prime p satisfying $n < p < 2n$ (having verified it for $n < 3\,000\,000$). He was not destined to provide a proof, but five years later the Russian mathematician Pafnuty Chebychev (1821–1894) was, and he was to come close to proving another of the great results of mathematics: the Prime Number Theorem, but more of that much later.

Firstly, it is clear that any number which can be represented as a finite decimal can be written as a fraction with denominator a power of 10. Ten is two times five and so the denominator is a power of two times five, and, after possible cancellation, of the form $2^\alpha 5^\beta$. To show that H_n is not a finite decimal, it is enough to show that the denominator of H_n, when it is written as a single fraction, contains prime factors greater than 5. Simply by writing out H_3, H_4 and H_5 we can establish that they are infinitely recurring; now write $H_n = a_n/b_n$, $n \geqslant 7$, where a_n and b_n are in their lowest terms. We need to show that b_n is divisible by some prime $p \geqslant 7$. To that end we will prove that for all primes $p \in [\frac{1}{2}(n+1), n]$, p divides b_n and do so by induction on n. For $n = 7$ the interval is $[4, 7]$, the set of primes $\{5, 7\}$ and since $H_7 = \frac{363}{140}$ we are done. Now assume the result for n, then we need to show that for all $p \in [\frac{1}{2}(n+2), n+1]$, p divides b_{n+1}, where

$$\frac{a_{n+1}}{b_{n+1}} = \frac{a_n}{b_n} + \frac{1}{n+1} = \frac{a_n(n+1) + b_n}{b_n(n+1)}.$$

Since this new interval can only add $n + 1$ to the list of primes and since if $n + 1$ is prime, $b_{n+1} = b_n(n+1)$ is incapable of cancellation with the a_{n+1}, we have what we need and the result is true by induction. The Bertrand Conjecture guarantees that the set of intervals $[p, 2p - 1]$ for $p \geqslant 7$ overlap and therefore contain every integer $n \geqslant 7$, since it guarantees a prime between every pair p and $2p$. Now we have all that we need. If $n \geqslant 7$, there is a prime $p \geqslant 7$ such that $n \in [p, 2p - 1]$, which means that $p \in [\frac{1}{2}(n+1), n]$ and so divides b_n. The reader may wish to look at what happens with $p = 5$.

Having studied the full harmonic series, we will look at some interesting subseries of it.

Sub-Harmonic Series

The mathematician requires tact and good taste at every step of his work, and he has to learn to trust to his own instinct to distinguish between what is really worthy of his efforts and what is not.

James Glaisher (1848–1928)

The incredibly slow divergence of H_n suggests that we would not need to alter its terms by much to force convergence, and by altering we mean omitting or cancelling. In this chapter, we will attempt just that.

3.1 A Gentle Start

If we start taking out terms in a structured way, we might start with

$$\frac{1}{2} + \frac{1}{4} + \frac{1}{6} + \frac{1}{8} + \cdots = \frac{1}{2}\left(1 + \frac{1}{2} + \frac{1}{3} + \frac{1}{4} + \cdots\right),$$

or alternatively

$$1 + \frac{1}{3} + \frac{1}{5} + \frac{1}{7} + \cdots > 1 + \frac{1}{4} + \frac{1}{6} + \frac{1}{8} + \cdots = 1 + \frac{1}{2}\left(\frac{1}{2} + \frac{1}{3} + \frac{1}{4} + \cdots\right),$$

both of which clearly diverge, which will have implications on p. 102.

So removing 'half' of the terms is not enough to force the depleted series to converge, nor would a third or any other fraction of it. Taking only powers of any single number leaves us with a convergent geometric series, but that really is taking out an awful lot of terms and not, in our development, very interesting. Is there something in between? A tantalizing possibility is to sum over the reciprocals of odd, perfect numbers (a perfect number is an integer which is equal to the sum of its proper divisors and 1; for example, 6 and 28). It is known that such a sum is finite; the problem is that no examples of these numbers are known and so our series may be entirely non-existent!

3.2 Harmonic Series of Primes

Primes are forever a source of interest and their scarcity (we will see just how scarce they are later) makes the series taken over the reciprocals of primes a pretty sparse one, and their lack of pattern (and we will be looking at that later too) a very attractive one.

$$\frac{1}{2} + \frac{1}{3} + \frac{1}{5} + \frac{1}{7} + \frac{1}{11} + \frac{1}{13} + \cdots$$

has indeed had a great deal removed from H_∞, but amazingly this also diverges. Of course, this must mean that there is an infinite number of primes, a fact established by Euclid in about 300 B.C. It is well worth a look at one version of his famous proof, as well as another entirely different, equally elegant and more modern argument. Firstly, Euclid.

Suppose that there are a finite number of primes and that the biggest of them is N, then the considerably bigger number composed of 1 plus the product of all of the primes up to and including N, $P = 1 + 2 \times 3 \times 5 \times 7 \times \cdots \times N$, either is prime (which would contradict our assumption that N is the biggest prime) or it is composite, and therefore divisible by primes. All of the primes leave a remainder of 1 when dividing P and so there must be other primes bigger than N, which is again a contradiction to the assumption that the number of primes is finite, with N as the biggest. The only escape is that the number of primes is infinite, and the proof is complete.

If P does happen to be prime it is given the appropriate name of 'Euclidean' prime. How common are these Euclidean primes? Things start off productively, with P a prime for N any of the first five primes 2, 3, 5, 7 and 11 (giving P as 3, 7, 31, 211 and 2311, respectively); the next Euclidean prime appears when $N = 31$, to give $P = 200\,560\,490\,131$, and the only other example for N less than 1000 is $N = 379$, with P rather too big to list! At present the largest known example is with $N = 24\,029$. Is there an infinite number? Nobody knows, but they do become very rare as N becomes very large.

A modern number theorist's proof of Euclid's result looks and feels different. In 1938 the consummate practitioner Paul Erdos (1913–1996) gave the one that follows, which uses a counting technique and a neat device used by number theorists: that any integer can always be written as the product of a square and a square-free integer. This is clear enough if the integer is factorized into the product of its prime factors and the repeated ones collected together; for example, $2\,851\,875 = 3^3 \times 5^4 \times 11 \times 13^2 = 3 \times 11 \times (3 \times 5^2 \times 13)^2$; of course, for a perfect square, the square-free part is 1. When we discuss the Riemann Hypothesis we will come across the Möbius function and see just how important it can be that an integer does or does not contain repeated factors. The proof is as follows.

Let N be any positive integer and $p_1, p_2, p_3, \ldots, p_n$ the complete set of primes less than or equal to N, then each of the positive integers less than or equal to N can, of course, be written as a product of powers of the p_i and, using the above observation, in the form $p_1^{e_1} p_2^{e_2} p_3^{e_3} \ldots p_n^{e_n} \times m^2$, where $e_i \in \{0, 1\}$, depending on whether a particular prime is present or not. Consequently, there are 2^n ways of choosing the square-free prime factorization and clearly $m^2 \leqslant N$ and so $m \leqslant \sqrt{N}$. This means that the integers less than or equal to N can be chosen in at most $2^n \times \sqrt{N}$ ways and therefore that $N \leqslant 2^n \times \sqrt{N}$, which makes $2^n \geqslant \sqrt{N}$ and $n \geqslant \frac{1}{2} \log_2 N$. Since N is unbounded, so must the number of primes be.

The proof leaves one breathless and wondering how anyone could ever have thought of it, but that was part of the genius of the man.

With the prime series now definitely infinite, we look to establishing its divergence. Euler (inevitably) attacked the problem and in doing so brought about an incredible result that spawned the whole subject of analytic number theory. Here we will give a proof based on Erdos's extension of his argument. Suppose that the series does converge. Then there must be a tail of the series which sums to less than $\frac{1}{2}$, that is, there must exist an i such that

$$\frac{1}{p_{i+1}} + \frac{1}{p_{i+2}} + \frac{1}{p_{i+3}} + \cdots < \frac{1}{2}.$$

Now let $N_i(x)$ be the number of positive integers less than x which are divisible by only the first i primes. If n is one of them, as before we can write $n = k \times m^2$, where k is a square-free number. Since there are precisely i primes that could divide k, $k = p_1^{\alpha_1} p_2^{\alpha_2} p_3^{\alpha_3} \ldots p_i^{\alpha_4}$, where $\alpha_r \in \{0, 1\}$ and so there are 2^i possible values for k, depending on whether a particular prime is present or not. Clearly, $m^2 \leqslant n < x$ and so m can be chosen in fewer than $\sqrt{x}$ ways, consequently, $N_i(x) < 2^i \sqrt{x}$. The number of positive integers less than x which are divisible by a prime p is at most x/p (consider $p, 2p, 3p, \ldots, np$, where $np \leqslant x$ and so $n \leqslant x/p$), therefore the number of positive integers less than x which are divisible by any prime other than the first i primes is at most

$$\frac{x}{p_{i+1}} + \frac{x}{p_{i+2}} + \frac{x}{p_{i+3}} + \cdots,$$

which is, of course, less than $\frac{1}{2}x$. But by definition this is $x - N_i(x)$, hence $x - N_i(x) < \frac{1}{2}x$ and $N_i(x) > \frac{1}{2}x$. Combining these two bounds we have $\frac{1}{2}x < N_i(x) < 2^i \sqrt{x}$ and hence $\frac{1}{2}x < 2^i \sqrt{x}$, which is true only for $x < 2^{2i+2}$. Take $x > 2^{2i+2}$ and we have our contradiction! And a perfectly beautiful one too.

So, the sum of the reciprocals of the primes diverges—but how slowly? *Very* slowly. For example,

$$\sum_{\substack{p < 1 \text{ million} \\ p \text{ prime}}} \frac{1}{p} = 2.887\,289\ldots$$

Our computer, which adds in a new term to the sum every 10^{-9} seconds, would, after 15 (American) billion years, have summed the series to a number just over 4. We will look at a form of Euler's proof later and that will also provide us with a measure of the rate of this glacially slow divergence.

As with the full harmonic series, even though the sum of the reciprocals of primes diverges, it manages to miss every integer. The proof is surprisingly easier than the one for the harmonic series. In fact, for any sequence of distinct primes $p_1, p_2, \ldots, p_m$, if

$$\frac{1}{p_1} + \frac{1}{p_2} + \frac{1}{p_3} + \cdots + \frac{1}{p_m} = n,$$

then

$$\frac{1}{p_1} = n - \frac{1}{p_2} - \frac{1}{p_3} - \cdots - \frac{1}{p_m} = \frac{a}{p_2 p_3 p_4 \cdots p_m}$$

for some integer a, hence $a p_1 = p_2 p_3 p_4 \cdots p_m$ and so $p_2 p_3 p_4 \cdots p_m$ is divisible by p_1, which is impossible.

Leaving only the primes fails to force convergence. If we pursue this thread, the most natural next step is to leave only the twin primes, that is, consecutive pairs of primes; it is customary (but not universal) to ignore 2 for this purpose and to count 5 twice, so the pairs are $(3,5), (5,7), (11,13), \ldots, (1019,1021), \ldots$, and these are incredibly sparse. In fact, it is not even known whether there is an infinite number of them and therefore if our series is infinite (this is called the Twin Primes Conjecture). It is interesting to note that the pair $(1019,1021)$ generate two Euclidean primes. Using only twin primes, all that is left of H_∞ is

$$\left(\frac{1}{3} + \frac{1}{5}\right) + \left(\frac{1}{5} + \frac{1}{7}\right) + \left(\frac{1}{11} + \frac{1}{13}\right) + \left(\frac{1}{17} + \frac{1}{19}\right) + \cdots.$$

Do we achieve convergence now? Finally, the answer is yes, but no one is sure to exactly what number; it is about $1.902\,160\,582\,4\ldots$ and is known as Brun's constant, after the Norwegian mathematician, Viggo Brun (1885–1978), who, in 1919, established the convergence. Not much is known about it, although its size is a strong indicator of just how sparse twin primes are. Thomas Nicely provided the above estimate in 1994 and in the process uncovered the infamous and much-publicized Intel Pentium division bug ('for a mathematician to get this much publicity, he would normally have to shoot someone'), which made itself apparent with the pair of twin primes $824\,633\,702\,441$ and $824\,633\,702\,443$. His announcement to the world was by a now famous email, which began:

> It appears that there is a bug in the floating point unit (numeric coprocessor) of many, and perhaps all, Pentium processors. In short, the Pentium FPU is returning erroneous values for certain division operations. For example, $1/824\,633\,702\,441.0$ is calculated incorrectly (all digits beyond the eighth significant digit are in error)...

On 17 January 1995 Intel announced a pre-tax charge of \$475 million against earnings, as the total cost associated with the replacement of the flawed chips.

Incidentally, it would have been very convenient had the series diverged, as that would have meant that there is an infinite number of twin primes and so resolved the Twin Primes Conjecture. (The reader may convince themselves that 5 is the only candidate for repetition by reasoning that any prime greater than 3 must be of the form $6n \pm 1$, any pair of twin primes must be $6n - 1$ and $6n + 1$ and therefore that a consecutive sequence of three is impossible beyond 3, 5, 7.)

3.3 THE KEMPNER SERIES

The most novel culling of the terms of the harmonic series has to be due to A. J. Kempner, who in 1914 considered what would happen if all terms are removed from it which have a particular digit appearing in their denominators. For example, if we choose the digit 7, we would exclude the terms with denominators such as 7, 27, 173, 33 779, etc. There are 10 such series, each resulting from the removal of one of the digits 0, 1, 2, ..., 9, and the first question which naturally arises is just what percentage of the terms of the series are we removing by the process? For example, if we remove all terms involving 0 we are left with

$$1 + \frac{1}{2} + \frac{1}{3} + \cdots + \frac{1}{9} + \frac{1}{11} + \cdots + \frac{1}{19} + \frac{1}{21} + \text{etc.}$$
$$+ \frac{1}{99} + \frac{1}{111} + \cdots + \frac{1}{119} + \frac{1}{121} + \text{etc.} + \frac{1}{999} + \cdots,$$

whereas if we remove all terms including 1 we are left with

$$\frac{1}{2} + \frac{1}{3} + \cdots + \frac{1}{9} + \frac{1}{20} + \frac{1}{22} + \frac{1}{23} + \cdots + \frac{1}{30} + \frac{1}{32} + \text{etc.}$$
$$+ \frac{1}{99} + \frac{1}{200} + \frac{1}{202} + \text{etc.} + \frac{1}{999} + \cdots.$$

Up to a given limit, we can count exactly how many terms have been removed by grouping the denominators of the terms by the number of digits they have in them, firstly assuming that we are removing 0 (see Table 3.1).

This means that when we have culled the denominators involving a 0 we are left with

$$9 + 9^2 + 9^3 + 9^4 + \cdots + 9^n = \frac{9(9^n - 1)}{9 - 1}$$
$$= \tfrac{9}{8}(9^n - 1) \text{ terms of the } 10^n - 1 \text{ possible.}$$

If we now perform the same analysis when we remove the digit 1 instead, we have Table 3.2.

Table 3.1. Removing the digit 0.

Denominator range	Number of allowed denominators
$1 \to 9$	9
$10 \to 99$	$9 \times 9 = 9^2$
$100 \to 999$	$9 \times 9 \times 9 = 9^3$
$1000 \to 9999$	$9 \times 9 \times 9 \times 9 = 9^4$
$\vdots$	$\vdots$
$10^{n-1} \to 10^n - 1$	9^n

Table 3.2. Removing the digit 1.

Denominator range	Number of allowed denominators
$1 \to 9$	8
$10 \to 99$	8×9
$100 \to 999$	$8 \times 9 \times 9 = 8 \times 9^2$
$1000 \to 9999$	$8 \times 9 \times 9 \times 9 = 8 \times 9^3$
$\vdots$	$\vdots$
$10^{n-1} \to 10^n - 1$	$8 \times 9^{n-1}$

The difference arises from the fact that 0 is now allowable but cannot be the first digit of any number. Now we are left with

$$8 + 8 \times 9 + 8 \times 9^2 + 8 \times 9^3 + \cdots + 8 \times 9^{n-1}$$
$$= 8 \frac{9^n - 1}{9 - 1}$$
$$= 9^n - 1 \text{ terms of the } 10^n - 1 \text{ possible.}$$

It is obvious that this last argument is valid for each of the other digits $2, \ldots, 9$ even though the actual sums (given they exist) will vary with the digit removed.

Looked at in a different way, with the digit 0 the fraction of terms that we have removed is

$$\frac{(10^n - 1) - \frac{9}{8}(9^n - 1)}{10^n - 1} = 1 - \frac{9}{8}\frac{9^n - 1}{10^n - 1} \xrightarrow[n \to \infty]{} 1 - 0 = 1$$

and with the other digits it is

$$\frac{(10^n - 1) - (9^n - 1)}{10^n - 1} = 1 - \frac{9^n - 1}{10^n - 1} \xrightarrow[n \to \infty]{} 1 - 0 = 1.$$

That is, asymptotically, we have removed 'almost all' terms! Put another way, we have the initially startling fact that almost all integers contain every possible digit. If we reflect on the number of digits that integers have as they get bigger, this is less surprising perhaps.

So, we really have removed a great many terms from the harmonic series and it should be no surprise that the depleted series do in fact converge. To see this, again we need to take separately the cases of the removed digit being 0 or otherwise. If we look back to Table 3.1 we have that the nine single-digit integers are each greater than or equal to 1, which makes the terms with those as denominators each less than or equal to 1, the 9^2 double-digit integers are each greater than or equal to 10, which makes the terms with those as denominators each less than or equal to $\frac{1}{10}$, etc., to give an upper bound for the sum of the series of

$$9 \times 1 + 9^2 \times \frac{1}{10} + 9^3 \times \frac{1}{10^2} + 9^4 \times \frac{1}{10^3} + \cdots$$

$$= 9\left(1 + \left(\frac{9}{10}\right) + \left(\frac{9}{10}\right)^2 + \left(\frac{9}{10}\right)^3 + \cdots\right)$$

$$= \frac{9}{1 - \frac{9}{10}} = 90.$$

The necessary changes for the other digits brings about an upper bound of

$$8 \times 1 + 8 \times 9 \times \frac{1}{10} + 8 \times 9^2 \times \frac{1}{10^2} + 8 \times 9^3 \times \frac{1}{10^3} + \cdots$$

$$= 8 + 8 \times \frac{9}{10} + 8 \times \left(\frac{9}{10}\right)^2 + 8 \times \left(\frac{9}{10}\right)^3 + \cdots$$

$$= 8\left(\frac{1}{1 - \frac{9}{10}}\right) = 80.$$

These are loose bounds but they do their job and show that the series do indeed converge. Of course, the slowness of the convergence hinders the computation of the exact sums, but R. Baillie has provided a method for summing the series with great accuracy and economy which resulted in Table 3.3, here given to five decimal places.

3.4 MADELUNG'S CONSTANTS

Finally, having omitted terms, we can take the alternative route and cancel them, most famously by considering the series $1 - \frac{1}{2} + \frac{1}{3} - \frac{1}{4} + \cdots$ to get the alternating harmonic series, which sums to $\ln 2$, which is of course a special case of that Newton–Mercator logarithmic series. A more intriguing alternative is to consider a more complicated modification to get $-\frac{4}{1} + \frac{4}{2} + \frac{4}{4} - \frac{8}{5} + \frac{4}{8} -$

Table 3.3. The Kempner-depleted harmonic sums.

Missing digit	Sum
0	23.103 44
1	16.176 96
2	19.257 35
3	20.569 87
4	21.327 46
5	21.834 60
6	22.205 59
7	22.493 47
8	22.726 36
9	22.920 67

$\cdots + \frac{12}{100} - \cdots + \frac{16}{500} - \cdots$, which may at first seem a touch arbitrary. The pattern is revealed when the series is written in sigma notation, to get

$$\sum_{i=1}^{\infty} (-1)^i \frac{r_2(i)}{i}, \tag{3.1}$$

where $r_2(i)$ is the number of ways of representing the integer i as the sum of two squares (including 0 and negative integers, so $4 = 0^2 + 2^2 = 0^2 + (-2)^2 = 2^2 + 0^2 = (-2)^2 + 0^2$). The missing terms (denominators 3, 6, 7, $\ldots$) come about because not all integers can be so expressed. Whether or not a particular integer is capable of being expressed as the sum of two squares was originally established by Euler, when in 1738 he published the result that a positive integer can be so expressed if and only if each of its prime factors of the form $4k + 3$ occurs as an even power.

It is hardly obvious, but the above series does converge and the limit is known to be $-\pi \ln 2$. Less obvious still is the fact that the series is connected with rock salt. The crystallographic structure of NaCl is that of a cubic lattice and the electrostatic potential at the origin caused by unit charges at those lattice points is, by definition,

$$M_3 = \sum_{i,j,k=-\infty}^{\infty} \frac{(-1)^{i+j+k}}{\sqrt{i^2 + j^2 + k^2}},$$

where not all three variables can be simultaneously zero. The series is a very delicate one, as we can see by considering the subseries of it with $k = 0$ and $i = j$, which brings about the infinite harmonic series once again and that of course diverges. The erratic behaviour of the series can be seen in Figure 3.1, the first of many bizarre graphs that we will consider.

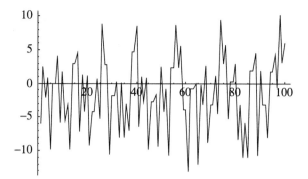

Figure 3.1. The electric potential of NaCl in three dimensions.

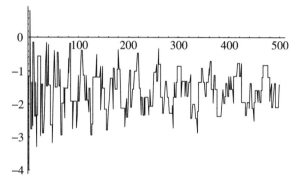

Figure 3.2. The electric potential of NaCl in two dimensions.

Notwithstanding this, a form of convergence can be defined for the series and with that definition of convergence its sum is $-1.747\,564\,59\ldots$, which is one of the Madelung constants. An alternative formulation is

$$\sum_{i=1}^{\infty} (-1)^i \frac{r_3(i)}{\sqrt{i}},$$

with $r_3(i)$ the number of ways in which the integer i can be written as the sum of three squares, which in Flatland reduces to

$$\sum_{i=1}^{\infty} (-1)^i \frac{r_2(i)}{\sqrt{i}},$$

with the cubic lattice becoming a square one and the convergence a much happier one, as Figure 3.2 indicates.

Its sum is $-1.615\,54\ldots$ and another Madelung constant (of which there is an infinite number as the dimension of space increases), and one which involves the Zeta function, which we will be meeting next.

Our series (3.1) is derived from this by omitting Rudolff's $\sqrt{}$ sign.

Zeta Functions

We may—paraphrasing the famous sentence of George Orwell—say that 'all mathematics is beautiful, yet some is more beautiful than the other'. But the most beautiful in all mathematics is the Zeta function. There is no doubt about it.

Krzysztof Maslanka

It is time to look at one of the 'advanced' functions of mathematics and one which lies at the core of the study of analytic number theory; a function which, according to M. C. Gutzwiller, 'is probably the most challenging and mysterious object of modern mathematics'. We will see it here in its own right and, in Chapter 6, linked to a second 'advanced' function and again in the final chapter, where its deepest behaviour is the stuff of the Riemann Hypothesis.

4.1 Where n Is a Positive Integer

The series

$$\sum_{r=1}^{\infty} \frac{1}{r^2} = 1 + \frac{1}{2^2} + \frac{1}{3^2} + \cdots$$

holds a special place in mathematical lore. A simple calculation suggests that it converges to the number $1.644\,934\ldots$, which is hardly illuminating, and as we have seen from the harmonic series, it might just be diverging very slowly. Actually, it does converge and is a special case of the whole family of convergent series defined for integers $n > 1$ by

$$\zeta(n) = \sum_{r=1}^{\infty} \frac{1}{r^n} = 1 + \frac{1}{2^n} + \frac{1}{3^n} + \cdots.$$

Bracketing the terms and comparing them with the geometric series establishes the convergence:

$$\sum_{r=1}^{\infty} \frac{1}{r^n} = 1 + \frac{1}{2^n} + \frac{1}{3^n} + \frac{1}{4^n} + \cdots$$

$$= 1 + \left(\frac{1}{2^n} + \frac{1}{3^n}\right) + \left(\frac{1}{4^n} + \frac{1}{5^n} + \frac{1}{6^n} + \frac{1}{7^n}\right) + \cdots$$

$$< 1 + \frac{2}{2^n} + \frac{4}{4^n} + \cdots$$

$$= 1 + \frac{1}{2^{n-1}} + \left(\frac{1}{2^{n-1}}\right)^2 + \left(\frac{1}{2^{n-1}}\right)^3 + \cdots$$

$$= \frac{1}{1 - \dfrac{1}{2^{n-1}}}$$

provided that $1/2^{n-1} < 1$, that is, $2^{n-1} > 1$, $n - 1 > 0$ and $n > 1$.

The above case (with $n = 2$) has a distinguished history, following its appearance in 1650, when Pietro Mengoli (1625–1686) asked for its value. John Wallis (1616–1703) computed it to three decimal places in 1665 but failed to recognize the significance of 1.645 (reasonably enough). In 1673 Oldenburg posed the problem to the great Gottfried von Leibnitz (1646–1716) (who was defeated by it) and it proved too much for other impressive mathematicians too, including Jacob Bernoulli (1654–1705), who had included a reference to it in his 1689 tract, published in Basel, *Tractatus de seriebus infinitis* with the entreaty, 'If anyone finds and communicates to us that which thus far has eluded our efforts, great will be our gratitude'; and so the problem has become known as the 'Basel Problem', 'the scourge of analysts', according to Montuela. The younger brother, Johann Bernoulli (1667–1748) (and mentor to the young Euler), tried and failed too and perhaps it was he who encouraged his brilliant student to attempt it—and having attempted it he eventually conquered it. In 1731 he computed the sum to six decimal places, in 1735 he sharpened his calculation to the number 1.644 934 066 848 226 436 47... and, later in that year, with his star still in its early ascendancy, he wrote, 'quite unexpectedly I have found an elegant formula involving the quadrature of the circle', by which he meant π. With his genius for analytic manipulation and his characteristic disregard for rigour he had shown that

$$\frac{1}{1^2} + \frac{1}{2^2} + \frac{1}{3^2} + \cdots = \frac{\pi^2}{6}.$$

The curious number 1.644 934... turns out to be $\frac{1}{6}\pi^2$, an astonishing result that did much to enhance Euler's growing reputation. Not unreasonably, it, combined with the divergence of the reciprocals of the primes, led him to remark (in 1737) that there are many more primes than perfect squares. It would take more than 100 years and the controversial work of another mathematical giant (Georg Cantor) to give rigour to this comment—and in doing so, to show it in that rigorous sense false.

Euler's original proof is magical and demands to appear here above all others, including a later, more careful and completely different version provided by him to answer critics. It begins with the standard Taylor expansion of $\sin x$,

$$\sin x = x - \frac{x^3}{3!} + \frac{x^5}{5!} - \frac{x^7}{7!} + \cdots,$$

which converges for all x. Euler interpreted the left-hand side as a polynomial of infinite degree. Since it is a polynomial it can be written as a product of factors and since the roots are $0, \pm\pi, \pm2\pi, \pm3\pi, \dots$, the polynomial can be written as

$$x(x^2 - \pi^2)(x^2 - 4\pi^2)(x^2 - 9\pi^2)\dots,$$

and this can be rewritten as

$$Ax\left(1 - \frac{x^2}{\pi^2}\right)\left(1 - \frac{x^2}{2^2\pi^2}\right)\left(1 - \frac{x^2}{3^2\pi^2}\right)\cdots$$

Since

$$\frac{\sin x}{x} \to 1 \quad \text{as } x \to 0,$$

it must be that $A = 1$. So,

$$\sin x = x - \frac{x^3}{3!} + \frac{x^5}{5!} - \frac{x^7}{7!} + \cdots = x\left(1 - \frac{x^2}{\pi^2}\right)\left(1 - \frac{x^2}{2^2\pi^2}\right)\left(1 - \frac{x^2}{3^2\pi^2}\right)\cdots$$

This astonishing piece of ingenuity is now part of the theory of infinite products, and through that theory is made rigorous. Now he equated the coefficients of x^3 on both sides to get

$$-\frac{1}{3!} = -\frac{1}{\pi^2} - \frac{1}{2^2\pi^2} - \frac{1}{3^2\pi^2} - \frac{1}{4^2\pi^2} - \cdots$$

or

$$\frac{1}{1^2} + \frac{1}{2^2} + \frac{1}{3^2} + \cdots = \frac{\pi^2}{6}$$

and the result has appeared as if from nowhere.

Bearing in mind the level of resourcefulness (and genius) required to establish the result, we can share A. G. Howson's amusement that 'one of the questions set to candidates for the first London University Matriculation Examination (in 1838), an examination set for students of 19 years or under who wished to enter the university, was: "Find the sum to infinity of the series

$$\frac{1}{1^2} + \frac{1}{2^2} + \frac{1}{3^2} + \cdots$$

and

$$\frac{1}{1 \times 2} + \frac{1}{2 \times 3} + \frac{1}{3 \times 4} + \cdots \text{"}$$

There is no indication how the examiner intended the question to be solved; the examination syllabus, which did not include the calculus, referred only to "arithmetical and geometrical progressions" and "arithmetic and algebra".' Partly through the connection with

$$\frac{1}{1^2} + \frac{1}{2^2} + \frac{1}{3^2} + \cdots ,$$

the number $\frac{1}{6}\pi^2$ turns up surprisingly often and frequently in unexpected places, as we shall see. A quite astonishing appearance of it is this: if we take two positive integers at random, the probability of them being co-prime (that is, having no common factors) is none other than 1 in $\frac{1}{6}\pi^2$. This is so shocking that we will take the considerable efforts that are needed to establish the proof, but before we can do so we will need more of Euler's help and so we must revisit it later.

In the final chapter we will mention three famous lists of mathematical problems, one of the turn of the 20th century, the second near its end and the third at the turn of the 21st century. Euler made four such. The first was read to the Mathematics Department of the University of Berlin on 6 September 1742 and consists of seven problems, not as a challenge to the mathematical community (as were the others) but as a list of ideas that he considered important and on which he was currently working. It was the following.

1. Determination of the orbit of the comet which was observed in the month of March in the year 1742.

2. Theorems about the reduction of integral formulas to the quadrature of circles.

3. On the finding of integrals which, if the value determined is assigned after the integration of the variable quantity.

4. On the sum of series of reciprocals arising from the powers of natural numbers.

5. On the integration of differential equations of higher degrees.

6. On the properties which certain conic sections have in common with infinitely many other curved lines.

7. On the resolution of the equations $dy + ayy\,dx = bxm\,dx$.

For the most part, they lack the crispness of the modern specification of a problem. Problem 3 seems obscure, problem 7 is a form of the Riccatti differential equation, which we would write as

$$\frac{dy}{dx} + ay^2 = bx^m,$$

and in problem 4 we see the reference to Zeta series.

There was ample evidence that his efforts were in part rewarded in his *Introductio* of 1748. In it, by equating other coefficients, he listed results for $\zeta(x)$ for $x = 2, 4, 6, \ldots, 26$. For example,

$$\zeta(4) = \frac{1}{1^4} + \frac{1}{2^4} + \frac{1}{3^4} + \cdots = \frac{\pi^4}{90}.$$

To demonstrate the difficulty of the problem, the sum for $x = 26$ is

$$\zeta(26) = \frac{1}{1^{26}} + \frac{1}{2^{26}} + \frac{1}{3^{26}} + \cdots$$

$$= \frac{2^{24} \times 76\,977\,927 \times \pi^{26}}{27!}$$

$$= \frac{1\,315\,862}{11\,094\,481\,976\,030\,578\,125} \pi^{26},$$

and all without a calculator.

Using similar ideas he was able to prove, for example,

$$\frac{1}{1^2} + \frac{1}{3^2} + \frac{1}{5^2} + \cdots = \frac{\pi^2}{8},$$

$$\frac{1}{1^4} + \frac{1}{3^4} + \frac{1}{5^4} + \cdots = \frac{\pi^4}{96},$$

$$\frac{1}{1^3} - \frac{1}{3^3} + \frac{1}{5^3} - \cdots = \frac{\pi^3}{32},$$

$$\frac{1}{1^5} - \frac{1}{3^5} + \frac{1}{5^5} - \cdots = \frac{5\pi^5}{1536}.$$

In a later paper, published in 1750, he recorded one of his major triumphs by solving the general problem for even n, showing that

$$\zeta(2n) = \sum_{r=1}^{\infty} \frac{1}{r^{2n}} = (-1)^{n-1} \frac{(2\pi)^{2n}}{2(2n)!} B_{2n},$$

where B_{2n} are the Bernoulli Numbers, which we will discuss in Chapter 10.

Astonishingly, no general formula is known for $\zeta(n)$ for n odd (and of course greater than 1), which makes the last two results listed above all the more tantalizing.

For interest, here are the first few sums to several decimal places:

$$\zeta(3) = \frac{1}{1^3} + \frac{1}{2^3} + \frac{1}{3^3} + \cdots = 1.202\,056\,903\,1\ldots,$$

$$\zeta(5) = \frac{1}{1^5} + \frac{1}{2^5} + \frac{1}{3^5} + \cdots = 1.036\,927\,755\,1\ldots,$$

$$\zeta(7) = \frac{1}{1^7} + \frac{1}{2^7} + \frac{1}{3^7} + \cdots = 1.008\,349\,277\,3\ldots.$$

$\zeta(3)$ is another of the many named mathematical constants; it is called Apery's constant, honouring Roger Apery, who, in 1978, proved it to be irrational. Not even that is known of any one of the others, even though the sums for even n are obviously transcendental. Given the pattern that exists for even powers, it is tempting to conjecture that

$$\zeta(2n+1) = \sum_{r=1}^{\infty} \frac{1}{r^{2n+1}} = \frac{p}{q} \pi^{2n+1}$$

for some integers p and q, which, in the case $n = 2$, would amount to trying to prove that

$$\frac{1.036\,927\,755\,1\ldots}{\pi^5} = 0.003\,388\,434\ldots$$

is rational. Umm. There is inexorable progress though. In 2000, T. Rival proved that there are infinitely many integers n such that $\zeta(2n+1)$ is irrational, and subsequently in 2001 that at least one of $\zeta(5), \zeta(7), \zeta(9), \ldots, \zeta(21)$ is irrational. Again in 2001 this result has been tightened by Zudilin to replace 21 by 11.

4.2 WHERE x IS A REAL NUMBER

We have been looking at $\zeta(n)$ for n a positive integer. The earlier proof showed that

$$\zeta(n) = \sum_{r=1}^{\infty} \frac{1}{r^n}, \quad n > 1,$$

is meaningful and made no assumption about n being an integer. If we replace n by the continuous, real variable $x > 1$, we meet the real 'Zeta function', whose graph is shown in Figure 4.1.

The vertical asymptote is at $x = 1$ because of the divergence of $\zeta(1)$ and the horizontal asymptote is at $y = 1$ since the terms of $\zeta(x)$ beyond the first contribute vanishingly small amounts as $x \to \infty$.

The asymptotic behaviour can be more exactly measured. If we overestimate the area under $y = 1/u^x$ for fixed x, between $u = 1$ and $u = n+1$ by rectangles

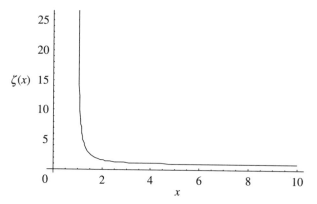

Figure 4.1. The Zeta function.

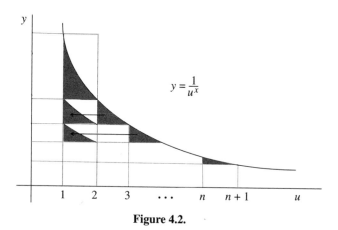

Figure 4.2.

of width 1, as in Figure 4.2, we have that

$$\left| \sum_{u=1}^{n} \frac{1}{u^x} - \int_{1}^{n+1} \frac{du}{u^x} \right| < 1,$$

since this quantity is just the sum of the areas of the shaded, curved triangles at the top of each region, which can be slid to the left to fit in the first rectangle, which has area 1. This means that

$$\left| \sum_{u=1}^{n} \frac{1}{u^x} - \frac{1}{x-1}\left(1 - \frac{1}{(n+1)^{x-1}} \right) \right| < 1$$

and so, as $n \to \infty$, $|\zeta(x) - 1/(x-1)| \leqslant 1$, which means that

$$|(x-1)\zeta(x) - 1| \leqslant |x-1|.$$

43

Now take the limit as $x \to 1^+$ and we have that $(x-1)\zeta(x) \to 1$ as $x \to 1^+$.

We will use this result later and also extend the definition of the Zeta function once again, this time from real x to complex z, with profound implications.

4.3 Two Results to End With

Before we move on to see how the Zeta functions bring about existence of Gamma (with a little help from Euler), we will mention two miscellaneous and nice results related to them.

Firstly, we know that the prime series

$$\sum_{p \text{ prime}} \frac{1}{p}$$

diverges, but it must be that the series of prime powers converges. Exactly to what numbers is yet another question with no answer but we can at least conclude that

$$\sum_{p \text{ prime}} \frac{1}{p^n} \leqslant \sum_{p \text{ prime}} \frac{1}{p^2} < \sum_{r=2}^{\infty} \frac{1}{r^2} = 1 - \frac{\pi^2}{6} < 1 \quad \text{for } n > 1,$$

which isn't much but it's about as much as we can expect for so little work in this most difficult area of mathematics.

The final item we will mention is a 1697 result of Johann Bernoulli, and is very easy on the eye. It is that

$$\int_0^1 \frac{1}{x^x}\,dx = \frac{1}{1^1} + \frac{1}{2^2} + \frac{1}{3^3} + \cdots .$$

The integral is improper, with 0^0 indeterminate, but we also have the well-known result that $\lim_{x \to 0} x^x = 1$ and with this in place we can indulge in a feast of integration by parts to prove the formula

$$\int_0^1 \frac{1}{x^x}\,dx = \int_0^1 e^{-x \ln x}\,dx = \int_0^1 \sum_{r=0}^{\infty} \frac{(-x \ln x)^r}{r!}\,dx$$

$$= \sum_{r=0}^{\infty} \frac{1}{r!} \int_0^1 (-x \ln x)^r\,dx = \sum_{r=0}^{\infty} \frac{(-1)^r}{r!} \int_0^1 x^r \ln^r x\,dx$$

$$= 1 + \sum_{r=1}^{\infty} \frac{(-1)^r}{r!} \int_0^1 x^r \ln^r x\,dx.$$

Now we attack the integral using parts and the fact that $\ln x$ grows much more slowly than any power of x to get

$$\int_0^1 x^r \ln^r x \, dx = \left[\frac{x^{r+1}}{r+1} \ln^r x\right]_0^1 - \frac{r}{r+1} \int_0^1 \frac{x^{r+1}}{x} \ln^{r-1} x \, dx$$

$$= -\frac{r}{r+1} \int_0^1 x^r \ln^{r-1} x \, dx$$

$$= \ldots (-1)^r \frac{r!}{(r+1)^r} \int_0^1 x^r \, dx$$

$$= (-1)^r \frac{r!}{(r+1)^{r+1}}$$

and so

$$\int_0^1 \frac{1}{x^x} \, dx = 1 + \sum_{r=1}^{\infty} \frac{(-1)^r}{r!} (-1)^r \frac{r!}{(r+1)^{r+1}}$$

$$= 1 + \sum_{r=1}^{\infty} \frac{1}{(r+1)^{r+1}}$$

$$= \frac{1}{1^1} + \frac{1}{2^2} + \frac{1}{3^3} + \cdots .$$

45

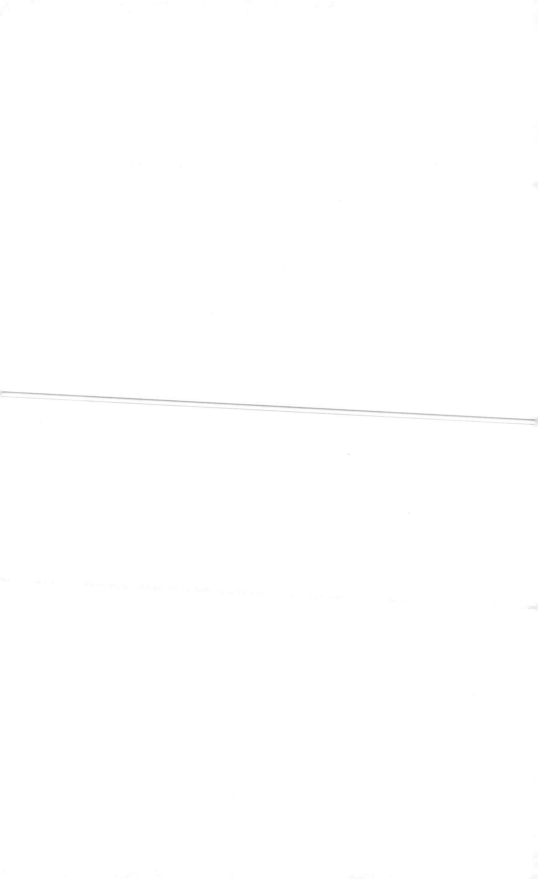

Gamma's Birthplace

The mathematician may be compared to a designer of garments, who is utterly oblivious of the creatures whom his garments may fit. To be sure, his art originated in the necessity for clothing such creatures, but this was long ago; to this day a shape will occasionally appear which will fit into the garment as if the garment had been made for it. Then there is no end of surprise and delight!

Tobias Dantzig

5.1 ADVENT

So, the harmonic series diverges, slowly. Just how slowly can be measured using its interpretation as a discrete logarithm. The area $\int_1^n (1/x)\,dx = \ln n$ is bounded below by the areas of the underestimating rectangles and above by the areas of the overestimating rectangles, which using Figures 5.1 and 5.2 results in the inequality

$$\frac{1}{2} + \frac{1}{3} + \cdots + \frac{1}{n} < \int_1^n \frac{1}{x}\,dx < 1 + \frac{1}{2} + \frac{1}{3} + \cdots + \frac{1}{n-1},$$

i.e.

$$H_n - 1 < \ln n < H_n - \frac{1}{n}$$

or

$$\ln n + \frac{1}{n} < H_n < \ln n + 1.$$

We have an estimate of H_n as $\ln n$ with an error of at least $1/n$ and at most 1, with H_n confined between the curves, as shown in Figure 5.3. Put another way,

$$\frac{1}{n} < H_n - \ln n < 1$$

and so, if the limit exists, $0 \leqslant \lim_{n\to\infty}(H_n - \ln n) \leqslant 1$.

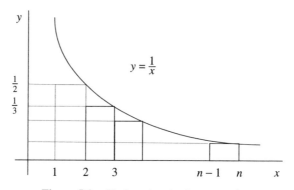

Figure 5.1. Underestimating by rectangles.

If we underestimate using trapezia, as in Figure 5.4, we achieve a different insight:

$$\int_1^n \frac{1}{x}\,dx$$

$$= \ln n$$

$$\approx \frac{1}{2}\left(1 + \frac{1}{2}\right) + \frac{1}{2}\left(\frac{1}{2} + \frac{1}{3}\right) + \frac{1}{2}\left(\frac{1}{3} + \frac{1}{4}\right) + \cdots + \frac{1}{2}\left(\frac{1}{n-1} + \frac{1}{n}\right)$$

$$= \frac{1}{2}\left(1 + \frac{1}{2} + \frac{1}{2} + \frac{1}{3} + \frac{1}{3} + \frac{1}{4} + \frac{1}{4} + \cdots + \frac{1}{n-1} + \frac{1}{n-1} + \frac{1}{n}\right)$$

$$= \frac{1}{2}\left(1 + 2\left(\frac{1}{2} + \frac{1}{3} + \frac{1}{4} + \cdots + \frac{1}{n-1}\right) + \frac{1}{n}\right)$$

$$= \frac{1}{2}\left(1 + 2\left(H_n - 1 - \frac{1}{n}\right) + \frac{1}{n}\right)$$

$$= \frac{1}{2}\left(2H_n - 1 - \frac{1}{n}\right)$$

$$= H_n - \frac{1}{2} - \frac{1}{2n}.$$

Therefore,

$$H_n \approx \ln n + \frac{1}{2} + \frac{1}{2n},$$

which means that

$$H_n - \ln n \approx \frac{1}{2} + \frac{1}{2n}$$

and so we may reasonably think that

$$\lim_{n\to\infty} (H_n - \ln n) \approx 0.5.$$

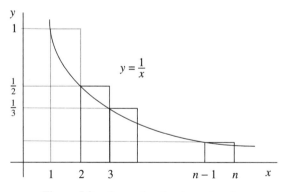

Figure 5.2. Overestimating by rectangles.

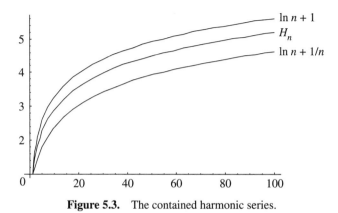

Figure 5.3. The contained harmonic series.

So, it looks like the difference between the harmonic series and the natural logarithm might tend to a number between 0 and 1 and near 0.5.

5.2 BIRTH

We have already mentioned that in 1735 Euler established the remarkable fact that

$$\zeta(2) = \frac{1}{1^2} + \frac{1}{2^2} + \frac{1}{3^2} + \cdots = \frac{\pi^2}{6},$$

and thereby solved the 'Basel problem', which had been frustrating mathematicians for years. In that same year, he published the paper 'De Progressionibus harmonicus observationes', which disclosed a further natural interest in Zeta functions and which led to Gamma coming into existence. We will look at the relevant part of the paper, using Euler's own invention of $\sum$, although he did

49

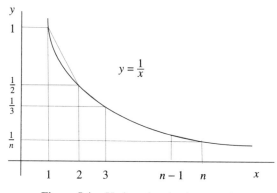

Figure 5.4. Underestimating by trapezia.

not make use of it himself on this occasion. Using the ubiquitous result

$$\ln(1 + x) = x - \tfrac{1}{2}x^2 + \tfrac{1}{3}x^3 - \tfrac{1}{4}x^4 + \cdots - 1 < x \leqslant 1,$$

he replaced x by $1/r$ to get

$$\ln\left(1 + \frac{1}{r}\right) = \frac{1}{r} - \frac{1}{2r^2} + \frac{1}{3r^3} - \frac{1}{4r^4} + \cdots$$

and so

$$\frac{1}{r} = \ln\left(\frac{r+1}{r}\right) + \frac{1}{2r^2} - \frac{1}{3r^3} + \frac{1}{4r^4} - \cdots$$

and

$$\sum_{r=1}^{n} \frac{1}{r} = \sum_{r=1}^{n} \ln\left(\frac{r+1}{r}\right) + \frac{1}{2}\sum_{r=1}^{n}\frac{1}{r^2} - \frac{1}{3}\sum_{r=1}^{n}\frac{1}{r^3} + \frac{1}{4}\sum_{r=1}^{n}\frac{1}{r^4} - \cdots;$$

therefore,

$$\sum_{r=1}^{n} \frac{1}{r} = \sum_{r=1}^{n}(\ln(r+1) - \ln r) + \frac{1}{2}\sum_{r=1}^{n}\frac{1}{r^2} - \frac{1}{3}\sum_{r=1}^{n}\frac{1}{r^3} + \frac{1}{4}\sum_{r=1}^{n}\frac{1}{r^4} - \cdots$$

and

$$\sum_{r=1}^{n} \frac{1}{r} = \ln(n+1) + \frac{1}{2}\sum_{r=1}^{n}\frac{1}{r^2} - \frac{1}{3}\sum_{r=1}^{n}\frac{1}{r^3} + \frac{1}{4}\sum_{r=1}^{n}\frac{1}{r^4} - \cdots,$$

which makes

$$\sum_{r=1}^{n} \frac{1}{r} - \ln(n+1) = \frac{1}{2}\sum_{r=1}^{n}\frac{1}{r^2} - \frac{1}{3}\sum_{r=1}^{n}\frac{1}{r^3} + \frac{1}{4}\sum_{r=1}^{n}\frac{1}{r^4} - \cdots.$$

In the limit as $n \to \infty$ we have the difference between the divergent harmonic series and the divergent natural logarithm expressed in terms of an infinite number of convergent Zeta series, the sums of which would therefore be very nice to know. We have seen that Euler did eventually solve the general problem of summing the Zeta series for even powers but also that the problem with odd powers remains open to this day and of course he was bound to resort to numeric methods to approximate the sum on the right-hand side, which in the 'De Progressionibus' he announced as 0.577 218.

In fact, Euler had the logarithm on the right-hand side to give

$$\sum_{r=1}^{n} \frac{1}{r} = \ln(n+1) + \frac{1}{2} \sum_{r=1}^{n} \frac{1}{r^2} - \frac{1}{3} \sum_{r=1}^{n} \frac{1}{r^3} + \frac{1}{4} \sum_{r=1}^{n} \frac{1}{r^4} - \cdots.$$

In his own words from 'De Progressionibus':

> Quae series cum sint convergentes, si proxime summentur prodibit
>
> $$1 + \frac{1}{2} + \cdots + \frac{1}{i} = \log(i+1) + 0.577\,218$$
>
> Si summa dicatur s, foret, ut supra fecimus,
>
> $$ds = \frac{di}{i+1},$$
>
> ideoque $s = \log(i+1) + C$. Hujus igitur quantitatis constantis C valorem deteximus, quippe est $C = 0.577\,218$.

Moving from 18th-century Latin to 21st-century English:

> This series, since each term is convergent taken one after the other, will proceed
>
> $$1 + \frac{1}{2} + \cdots + \frac{1}{i} = \log(i+1) + 0.577\,218.$$
>
> If the sum is called s it would be that
>
> $$ds = \frac{di}{i+1}$$
>
> as we have seen above, and so $s = \log(i+1) + C$. Therefore, we have revealed the value of this constant to this accuracy to be $C = 0.577\,218$.

And a birth is recorded under the name of C. Other letters have subsequently been used but it is γ that has become permanently attached to the number which, as we mentioned in the Introduction, he regarded as 'worthy of serious consideration'. He lavished considerable attention on it himself, partly hoping to identify it in terms of some other known constant or function. In 1781 under the name of C (albeit with the logarithm of n rather than $n+1$) he communicated the memoir 'De Numero Memorabili in Summatione Progessionis Harmonicae Naturalis Occurente' to the Petersburg Academy, which was entirely devoted to its study, and in which he admits that its nature still eluded him. He remarked that he had hoped that his C was itself the logarithm of another number of import but, having failed to identify any such number, continued by giving a whole list of series by which its approximate value might be calculated, two of which were

$$\sum_{i=2}^{\infty} \frac{1}{i}(\zeta(i) - 1) = 1 - \gamma$$

and

$$\sum_{i=1}^{\infty} \frac{1}{(2i + 1)2^{2i}}(\zeta(2i) - 1) = 1 - \gamma - \ln \tfrac{3}{2}.$$

He used the first (which we will prove in Chapter 12) to evaluate the constant to five decimal places and the second to evaluate it to 12 decimal places of accuracy.

The years have passed and the number has indeed been afforded that 'serious consideration' by any number of mathematicians but has hardly cooperated, and even at its venerable age of 267+ it is still so deeply shrouded in mystery that it is not even known if it is a fraction. In fact, the great G. H. Hardy, whom we will soon discuss, offered to vacate his Savilian Chair at Oxford to anyone who could prove Gamma to be irrational!

The Gamma Function

There is no branch of mathematics, however abstract, which may not some day be applied to phenomena of the real world.

Nikolai Lobatchevsky (1792–1856)

We will now look at that second 'advanced' function and its link with Euler's constant and with the Zeta function.

6.1 EXOTIC DEFINITIONS...

The striking integral

$$\int_0^1 \left(\ln \left(\frac{1}{t} \right) \right)^{x-1} dt$$

occupied some of Euler's many mathematical thoughts during the years 1729 and 1730 and in a letter to Christian Goldbach (1690–1764), dated 8 January 1730, he proposed its use in a quite startling way. It converges for $x > 0$ and can be considered as a function of x in that domain, a function whose properties are surprising and unexpectedly useful. In 1809 Adrien-Marie Legendre (1752–1833) gave it the name Gamma and the matching symbol Γ and so we have

$$\Gamma(x) = \int_0^1 \left(\ln \left(\frac{1}{t} \right) \right)^{x-1} dt = \int_0^1 (-\ln t)^{x-1} dt, \quad x > 0.$$

The substitution $t \to -\ln t$ results in the useful alternative

$$\Gamma(x) = \int_0^\infty t^{x-1} e^{-t} dt, \quad x > 0.$$

Clearly,

$$\Gamma(1) = \int_0^\infty e^{-t} dt = [-e^{-t}]_0^\infty = 1.$$

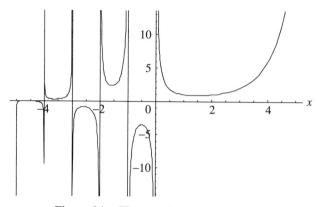

Figure 6.1. The extended Gamma function.

And also

$$\Gamma(x+1) = \int_0^\infty t^x e^{-t}\, dt$$

$$= [-t^x e^{-t}]_0^\infty + x\int_0^\infty t^{x-1} e^{-t}\, dt = x\Gamma(x).$$

This last property is its 'functional relationship', which can be used to extend the definition beyond $x > 0$ (we will meet a further critically important functional relationship in the final chapter) by rewriting the identity as

$$\Gamma(x) = \frac{\Gamma(x+1)}{x}$$

and so, for example, $\Gamma(-\frac{1}{2}) = -2\Gamma(\frac{1}{2})$. The vertical asymptote at $x = 0$ prevents the function being meaningful for the negative integers but otherwise the extension is to all $\mathbb{R}$ (and later to $\mathbb{C}$, minus those integers). Its graph is shown in Figure 6.1.

The function begins to reveal some of its subtleties when we take $x = n$ to be a positive integer, since the functional relationship becomes

$$\Gamma(n) = (n-1)\Gamma(n-1) = (n-1)(n-2)\Gamma(n-2)$$
$$= (n-1)(n-2)(n-3)\Gamma(n-3) = \cdots = (n-1)!$$

and so the Gamma function can be thought of as an extension of the factorial function, which is defined only for positive integers. If we allow the exclamation mark to be used in this extended sense (rather than using Γ) we discover the painful 'factorial fact', disbelieved by so many students, that

$$0! = (1-1)! = \Gamma(1) = 1.$$

If we accept a standard result that

$$\int_0^\infty e^{-u^2}\, du = \frac{\sqrt{\pi}}{2},$$

we can easily develop other exotic looking things such as

$$(\tfrac{1}{2})! = \Gamma(\tfrac{3}{2}) = \tfrac{1}{2}\Gamma(\tfrac{1}{2}) = \frac{1}{2}\int_0^\infty t^{-1/2}e^{-t}\, dt = \int_0^\infty e^{-u^2}\, du = \frac{\sqrt{\pi}}{2}$$

and the possibly even more striking

$$(-\tfrac{1}{2})! = \Gamma(\tfrac{1}{2}) = \sqrt{\pi}.$$

Of course, infinitely many exact values of Γ can be generated in this way, but interestingly there is no known exact value for $\Gamma(\tfrac{1}{3})$ or $\Gamma(\tfrac{1}{4})$ or infinitely many other values, although many are known to be transcendental.

In fact, on 13 October 1729, Euler had already proposed to Goldbach the definition

$$\Gamma(x) = \lim_{r \to \infty} \Gamma_r(x),$$

where

$$\Gamma_r(x) = \frac{r!\, r^x}{x(1+x)(2+x)\cdots(r+x)}$$

$$= \frac{r^x}{x\left(1+\dfrac{x}{1}\right)\left(1+\dfrac{x}{2}\right)\cdots\left(1+\dfrac{x}{r}\right)},$$

and for the moment this turns out to be a more useful form than the previous two.

It is hardly obvious that this is in fact the Gamma function, but we can recover the original definition by establishing that in the limit the functional relationship and boundary condition are satisfied,

$$\Gamma_r(x+1) = \frac{r!\, r^{x+1}}{(x+1)(x+2)\cdots(x+r)(x+1+r)}$$

$$= \frac{r}{x+r+1}x\Gamma_r(x)$$

so

$$\Gamma(x+1) = \lim_{r \to \infty} \Gamma_r(x+1) = \lim_{r \to \infty} \frac{r}{x+r+1}x\Gamma_r(x) = x\Gamma(x),$$

which is the functional relationship and

$$\Gamma_r(1) = \frac{r!}{(1+1)(1+2)\cdots(1+r)}r = \frac{r}{r+1}$$

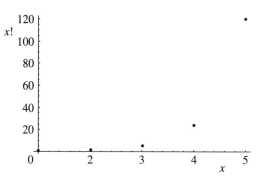

Figure 6.2. The factorial function.

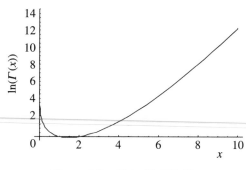

Figure 6.3. Plot of $\ln(\Gamma(x))$.

so

$$\Gamma(1) = \lim_{r \to \infty} \Gamma_r(1) = \lim_{r \to \infty} \frac{r}{r+1} = 1$$

and the boundary condition is indeed satisfied.

6.2 ... YET REASONABLE DEFINITIONS

This all might seem a bit contrived. Why generalize the factorial in such a seemingly bizarre way? After all, if we think about the problem geometrically we have the discrete factorial as in Figure 6.2 and what we want to do is to join the dots in a useful way. However we join them, we will want an explicit formula and if we write the extension as $f(x)$, then certainly we want $f(1) = 1$ and $f(x+1) = xf(x)$. Do these conditions restrict us to a single way of joining up those dots? The answer is 'no' but we need only one more reasonable condition to change that answer to 'yes' and that condition is pointed to by a significant result of 1922, known as the Bohr–Mollerup Theorem. If we look at the plot of $\ln(\Gamma(x))$ for $x > 0$, as shown in Figure 6.3, we see that it is always convex. The Bohr–Mollerup Theorem tells us that, with the two conditions above and with $\ln(f(x))$ convex, $f(x)$ must be the Gamma function—no other function will do!

6.3 GAMMA MEETS GAMMA

Karl Weierstrass (1815–1897) rewrote the definition and by so doing brought about the link between Gamma the number and Gamma the function,

$$\Gamma_r(x) = \frac{e^{x \ln r}}{x\left(1 + \dfrac{x}{1}\right)\left(1 + \dfrac{x}{2}\right) \cdots \left(1 + \dfrac{x}{r}\right)}$$

$$= \frac{e^{x(\ln r - 1 - 1/2 - 1/3 - \cdots - 1/r)} e^{(x + x/2 + x/3 + \cdots + x/r)}}{x\left(1 + \dfrac{x}{1}\right)\left(1 + \dfrac{x}{2}\right) \cdots \left(1 + \dfrac{x}{r}\right)}$$

$$= e^{x(\ln r - 1 - 1/2 - 1/3 - \cdots - 1/r)}$$

$$\times \frac{1}{x} \frac{e^x}{\left(1 + \dfrac{x}{1}\right)} \frac{e^{x/2}}{\left(1 + \dfrac{x}{2}\right)} \frac{e^{x/3}}{\left(1 + \dfrac{x}{3}\right)} \cdots \frac{e^{x/r}}{\left(1 + \dfrac{x}{r}\right)}$$

$$= \frac{e^{-x(1 + 1/2 + 1/3 + \cdots + 1/r - \ln r)}}{x}$$

$$\times \frac{e^x}{\left(1 + \dfrac{x}{1}\right)} \frac{e^{x/2}}{\left(1 + \dfrac{x}{2}\right)} \frac{e^{x/3}}{\left(1 + \dfrac{x}{3}\right)} \cdots \frac{e^{x/r}}{\left(1 + \dfrac{x}{r}\right)}$$

and so

$$\frac{1}{\Gamma(x)} = \lim_{r \to \infty} \frac{1}{\Gamma_r(x)} = x e^{\gamma x} \prod_{r=1}^{\infty} \left(1 + \frac{x}{r}\right) e^{-x/r}$$

with

$$\lim_{r \to \infty} \left(\frac{1}{1} + \frac{1}{2} + \frac{1}{3} + \cdots + \frac{1}{r} - \ln r\right) = \lim_{r \to \infty} (H_r - \ln r) = \gamma.$$

If we take this a bit further, we have

$$-\ln \Gamma(x) = \ln x + \gamma x + \sum_{r=1}^{\infty} \left(\ln\left(1 + \frac{x}{r}\right) - \frac{x}{r}\right).$$

Differentiating both sides with respect to x and moving the minus sign across gives

$$\frac{\Gamma'(x)}{\Gamma(x)} = -\frac{1}{x} - \gamma + \sum_{r=1}^{\infty}\left(\frac{1}{r} - \frac{1/r}{1+x/r}\right)$$
$$= -\frac{1}{x} - \gamma + \sum_{r=1}^{\infty}\left(\frac{1}{r} - \frac{1}{r+x}\right),$$

which defines the Digamma (or Psi) function $\Psi(x) = \Gamma'(x)/\Gamma(x)$.

Now, evaluating this at $x = 1$ gives

$$\Psi(1) = \frac{\Gamma'(1)}{\Gamma(1)} = -1 - \gamma + \sum_{r=1}^{\infty}\left(\frac{1}{r} - \frac{1}{r+1}\right) = -\gamma$$

and so $\Gamma'(1) = -\gamma$, and we have the geometrically appealing result that γ is numerically the gradient of the Gamma function at the point with x-coordinate 1.

If we incorporate the $1/x$ into the sum, we have that

$$\Psi(x) = -\gamma + \sum_{r=1}^{\infty}\left(\frac{1}{r} - \frac{1}{r+x-1}\right)$$

and so

$$\Psi(x+1) - \Psi(x) = \sum_{r=1}^{\infty}\left(\frac{1}{r} - \frac{1}{r+x}\right) - \sum_{r=1}^{\infty}\left(\frac{1}{r} - \frac{1}{r+x-1}\right) = \frac{1}{x}$$

and we have a familiar looking recurrence relation

$$\Psi(x+1) = \Psi(x) + \frac{1}{x},$$

familiar because we can recall the recurrence definition of the harmonic series as

$$H_r = H_{r-1} + \frac{1}{r}$$

for $r > 1$, with $H_1 = 1$. Taking x as the non-negative integer n, using the condition $\Psi(1) = -\gamma$ and chasing the recurrence relation down those integers results in the nice relationship $\Psi(n) = -\gamma + H_{n-1}$.

6.4 COMPLEMENT AND BEAUTY

In this final section we will establish an important formula involving the Gamma function (again originally discovered by Euler), and a beautiful and far-reaching connection between it and the Zeta function.

Using the earlier result we can write

$$\frac{1}{\Gamma(x)} \frac{1}{\Gamma(-x)} = -x^2 e^{\gamma x} e^{-\gamma x} \prod_{r=1}^{\infty} \left(1 + \frac{x}{r}\right)\left(1 - \frac{x}{r}\right) e^{x/r} e^{-x/r},$$

but $\Gamma(1 - x) = -x\Gamma(-x)$ and so

$$\frac{1}{\Gamma(x)} \frac{1}{\Gamma(1 - x)} = x \prod_{r=1}^{\infty} \left(1 - \frac{x^2}{r^2}\right)$$

and since we have the magical Euler formula

$$\sin(\pi x) = \pi x \left(1 - \frac{x^2}{1^2}\right)\left(1 - \frac{x^2}{2^2}\right)\left(1 - \frac{x^2}{3^2}\right) \cdots$$

we have that

$$\frac{1}{\Gamma(x)} \frac{1}{\Gamma(1 - x)} = \frac{\sin(\pi x)}{\pi}$$

or the

Complement Formula

$$\Gamma(x)\Gamma(1 - x) = \frac{\pi}{\sin(\pi x)}$$

which is valid whenever x and $1 - x$ are not zero or negative integers. A 'reflection formula' for a function $f(x)$ is one which relates $f(x)$ to $f(a - x)$ for some constant a. The Complement Formula is then the reflection formula of the Gamma function, with $a = 1$.

Now recall that

$$\Gamma(x) = \int_0^{\infty} t^{x-1} e^{-t} \, dt \quad \text{for } x > 0$$

and make the change of variable $t = ru$ to get

$$\Gamma(x) = \int_0^{\infty} (ru)^{x-1} e^{-ru} r \, du = r^x \int_0^{\infty} u^{x-1} e^{-ru} \, du.$$

Hence

$$\frac{1}{r^x} = \frac{1}{\Gamma(x)} \int_0^{\infty} u^{x-1} e^{-ru} \, du$$

and

$$\zeta(x) = \sum_{r=1}^{\infty} \frac{1}{r^x} = \frac{1}{\Gamma(x)} \sum_{r=1}^{\infty} \int_0^{\infty} u^{x-1} e^{-ru} \, du$$

$$= \frac{1}{\Gamma(x)} \int_0^{\infty} u^{x-1} \sum_{r=1}^{\infty} e^{-ru} \, du,$$

having pushed the sigma through the integral, and summing the infinite geometric series results in

$$\zeta(x) = \frac{1}{\Gamma(x)} \int_0^\infty u^{x-1} \frac{e^{-u}}{1 - e^{-u}} \, du$$

so we get

A Beautiful Formula
$$\zeta(x)\Gamma(x) = \int_0^\infty \frac{u^{x-1}}{e^u - 1} \, du$$

which is valid for $x \notin \{1, 0, -1, -2, \ldots\}$; a relationship which we will later see has far-reaching consequences.

Euler's Wonderful Identity

In great mathematics there is a very high degree of unexpectedness, combined with inevitability and economy.

G. H. Hardy (1877–1947)

7.1 THE ALL-IMPORTANT FORMULA...

Euler wanted to establish the divergence of the reciprocals of the primes. We have already seen Erdos's stylish proof of this but that will not prevent us from revelling in the glory of Euler's inventiveness, particularly as it brought about a result which is the cornerstone of analytic number theory, and which we will have considerable use of later.

The positive integers are a Unique Factorization Domain, that is, every positive integer is uniquely expressible as a product of primes (which is why 1 is not considered prime), and from this innocent fact Euler extracted wonder by producing the equivalent of the following arguments.

Since for any positive integer r, we can write $r = 2^{r_1} 3^{r_2} 5^{r_3} \cdots$ for some $r_1, r_2, r_3, \ldots \in \{0, 1, 2, 3, \ldots\}$ we have that

$$\frac{1}{r^x} = \frac{1}{(2^{r_1} 3^{r_2} 5^{r_3} \cdots)^x} = \frac{1}{2^{x r_1} 3^{x r_2} 5^{x r_3} \cdots}$$

and for $x > 1$

$$\zeta(x) = \sum_{r=1}^{\infty} \frac{1}{r^x} = \sum_{r_1, r_2, r_3, \ldots \geq 0} \frac{1}{2^{x r_1} 3^{x r_2} 5^{x r_3} \cdots}$$

$$= \left(\sum_{r_1 \geq 0} \frac{1}{2^{x r_1}} \right) \left(\sum_{r_2 \geq 0} \frac{1}{3^{x r_2}} \right) \left(\sum_{r_3 \geq 0} \frac{1}{5^{x r_3}} \right) \cdots$$

$$= \prod_{p \text{ prime}} \left(\sum_{\alpha=0}^{\infty} \frac{1}{p^{x \alpha}} \right) = \prod_{p \text{ prime}} \left(\sum_{\alpha=0}^{\infty} \frac{1}{(p^x)^{\alpha}} \right).$$

Now each term is a geometric series summing to

$$\frac{1}{1 - 1/p^x} = \frac{1}{1 - p^{-x}},$$

which means that we have

> ### Euler's Formula
>
> $$\zeta(x) = \sum_{r=1}^{\infty} \frac{1}{r^x} = \prod_{p \text{ prime}} \frac{1}{1 - p^{-x}}, \quad x > 1.$$

With this result the primes, the building blocks of the integers, are inextricably linked with the Zeta functions and through this link analytic number theory came into being.

7.2 ... AND A HINT OF ITS USEFULNESS

We have already seen proofs of the infinity of the primes, but Euler's result quickly provides two more. Taking the limit as $x \to 1$ results in

$$\sum_{r=1}^{\infty} \frac{1}{r} = \prod_{p \text{ prime}} \frac{1}{1 - p^{-1}},$$

with the divergence of the harmonic series forcing the product to be infinite and therefore so must be the number of primes.

And, with the result for $\zeta(2)$, we have

$$\frac{\pi^2}{6} = \zeta(2) = \prod_{p \text{ prime}} \frac{1}{1 - p^{-2}},$$

with the right-hand side rational if there were to be a finite number of primes; since π^2 is irrational (proved by Legendre in 1796) it must be that there are an infinity of primes—once again.

Following Erdos's proof by contradiction that

$$\sum_{p \text{ prime}} \frac{1}{p}$$

is divergent, we can now taste the flavour of an Eulerian approach and also use it to give a useful estimate of the size of

$$\sum_{p} \frac{1}{p^x}.$$

To do this, take logarithms of Euler's identity to get

$$\ln \zeta(x) = \sum_{p \text{ prime}} \ln \left(1 - \frac{1}{p^x}\right)^{-1}.$$

Now apply that most useful logarithmic series, $\ln(1-t) = -t - \frac{1}{2}t^2 - \frac{1}{3}t^3 - \cdots$ with $t = 1/p^x$ to get

$$\ln \left(1 - \frac{1}{p^x}\right)^{-1} = \frac{1}{p^x} + \left(\frac{1}{2p^{2x}} + \frac{1}{3p^{3x}} + \frac{1}{4p^{4x}} + \cdots\right)$$

and

$$\ln \zeta(x) = \sum_{p \text{ prime}} \ln \left(1 - \frac{1}{p^x}\right)^{-1}$$

$$= \sum_{p \text{ prime}} \left[\frac{1}{p^x} + \left(\frac{1}{2p^{2x}} + \frac{1}{3p^{3x}} + \frac{1}{4p^{4x}} + \cdots\right)\right]$$

$$= \sum_{p \text{ prime}} \frac{1}{p^x} + \sum_{p \text{ prime}} \left(\frac{1}{2p^{2x}} + \frac{1}{3p^{3x}} + \frac{1}{4p^{4x}} + \cdots\right),$$

where

$$\frac{1}{2p^{2x}} + \frac{1}{3p^{3x}} + \frac{1}{4p^{4x}} + \cdots < \frac{1}{2p^{2x}} + \frac{1}{2p^{3x}} + \frac{1}{2p^{4x}} + \cdots$$

$$= \frac{1}{2p^{2x}} \left(1 + \frac{1}{p^x} + \left(\frac{1}{p^x}\right)^2 + \left(\frac{1}{p^x}\right)^3 + \cdots\right)$$

$$= \frac{1}{2p^{2x}} \left(1 - \frac{1}{p^x}\right)^{-1}.$$

Now, playing see-saw with the inequality signs, $p^x > 2$, so

$$\frac{1}{p^x} < \frac{1}{2} \quad \text{and} \quad 1 - \frac{1}{p^x} > 1 - \frac{1}{2} \quad \text{and} \quad \left(1 - \frac{1}{p^x}\right)^{-1} < \left(1 - \frac{1}{2}\right)^{-1} = 2,$$

which makes

$$\frac{1}{2p^{2x}} + \frac{1}{3p^{3x}} + \frac{1}{4p^{4x}} + \cdots < \frac{1}{p^{2x}}$$

and so

$$\sum_{p \text{ prime}} \left(\frac{1}{2p^{2x}} + \frac{1}{3p^{3x}} + \frac{1}{4p^{4x}} + \cdots\right) < \sum_{p \text{ prime}} \frac{1}{p^{2x}}$$

$$< \sum_{p \text{ prime}} \frac{1}{p^2} < \sum_{n} \frac{1}{n^2} = \zeta(2).$$

This means that

$$\ln \zeta(x) = \sum_{p \text{ prime}} \frac{1}{p^x} + \text{error},$$

where the error is less than $\zeta(2)$. But

$$\ln \zeta(x) = \ln[(x-1)\zeta(x)] + \ln \frac{1}{x-1}$$

and so

$$\ln[(x-1)\zeta(x)] + \ln \frac{1}{x-1} = \sum_{p \text{ prime}} \frac{1}{p^x} + \text{error}.$$

Recalling the result from p. 44 we have that, for all $x > 1$,

$$\sum_{p \text{ prime}} \frac{1}{p^x} = \ln \frac{1}{x-1} + \text{bounded error}.$$

Now let $x \to 1$ and we have the divergence.

We can estimate the rate of divergence by saying that for large n,

$$\prod_{p<n} \frac{1}{1-p^{-1}} \approx \sum_{r<n} \frac{1}{r} \approx \ln n$$

and taking logs gives

$$-\sum_{p<n} \ln(1-p^{-1}) \approx \ln \ln n,$$

which means that

$$-\sum_{p<n} \left(-\frac{1}{p} - \frac{1}{2p^2} - \cdots \right) \approx \ln \ln n$$

and so

$$\sum_{p<n} \frac{1}{p} \approx \ln \ln n.$$

The reciprocals of the primes diverge as an approximate double ln. A more careful (and rigorous) argument shows that

$$\lim_{n\to\infty} \left(\sum_{p<n} \frac{1}{p} - \ln \ln n \right) = \gamma + \sum_{p \text{ prime}} \left(\ln \left(1 - \frac{1}{p}\right) + \frac{1}{p} \right)$$

$$= 0.261\,497\,212\,8\ldots$$

with another reappearance of γ and an appearance of one of the Meissel–Mertens constants.

Later, we will have considerably more work for Euler's formula to do.

A Promise Fulfilled

The good Christian should beware of mathematicians, and all those who make empty prophesies. The danger already exists that the mathematicians have made a covenant with the devil to darken the spirit and to confine man in the bonds of Hell.

St Augustine (354–430)[1]

Earlier we mentioned the barely credible result that the probability of two randomly chosen integers being co-prime is $1:\frac{1}{6}\pi^2$. With Euler's formula, combined with several other mathematical tools listed below, we are able to prove the fact; but first those tools.

(1) In set theory, the symbols $\cap$ and $\cup$ (respectively, the intersection and union of sets) are defined to be the set of all elements common to both and contained in either or both, respectively. These 'binary' operations on sets give rise to an algebra, known as 'Boolean algebra', named after the English mathematician George Boole (1815–1864), from which we need only the distributive law $A \cap (B \cup C) = (A \cap B) \cup (A \cap C)$. (Incidentally, the reader in search of greater challenge than this book can offer might wish to consult G. Spencer-Brown's 1969 publication *Laws of Form*, in which he develops an arithmetic for Boolean algebra.)

If $n(A)$ is taken to mean the number of elements in the set A, we can easily see that $n(A \cup B) = n(A) + n(B) - n(A \cap B)$ and, using the distributive law, that

$$
\begin{aligned}
n(A \cup B \cup C) &= n(A \cup (B \cup C)) \\
&= n(A) + n(B \cup C) - n(A \cap (B \cup C)) \\
&= n(A) + n(B) + n(C) - n(B \cap C) - n((A \cap B) \cup (A \cap C)) \\
&= n(A) + n(B) + n(C) - n(B \cap C) - n(A \cap B) - n(A \cap C) \\
&\quad + n(A \cap B \cap C).
\end{aligned}
$$

---

[1] Here, 'mathematician' means 'astrologer'.

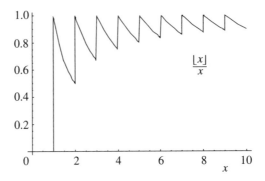

Figure 8.1. The Floor function compared with x.

Using induction or otherwise, it is easy to prove that the pattern of 'one at a time minus two at a time plus three at a time minus four at a time, etc.' continues to any number of sets. This result is often called the *inclusion–exclusion principle*.

(2) An equivalent result. The expression $(1-x_1)(1-x_2)(1-x_3)(1-x_4)\cdots$ when expanded takes the form

$$1 - (x_1 + x_2 + x_3 + x_4 + \cdots) + (x_1x_2 + x_2x_3 + x_3x_4 + \cdots)$$
$$- (x_1x_2x_3 + x_2x_3x_4 + x_1x_2x_4 + \cdots) + \cdots,$$

where the brackets contain the sums of the x taken one at a time, two at a time, three at a time, etc., and the signs between them alternate.

(3) The modern form of the Greatest Integer function, $[\,\cdot\,]$, are the Floor and Ceiling functions, succeeded in name and notation in the 1960s when Kenneth E. Iverson introduced them. The definitions are, respectively,

$\lfloor x \rfloor$ is the greatest integer $\leqslant x$ and $\lceil x \rceil$ is the smallest integer $\geqslant x$.

If N and n are positive integers with $n \leqslant N$ and the sequence $1n, 2n, 3n, \ldots,$ xn stops where x is the biggest multiple such that $xn \leqslant N$, then $x = \lfloor N/n \rfloor$. This means that there are $\lfloor N/n \rfloor$ numbers up to and including N that have n as a divisor. We will use this fact on several occasions throughout the book, and with Erdos's proof on p. 29 have already done so.

Also notice that $\lfloor x \rfloor = x - \alpha$ for $0 \leqslant \alpha < 1$, which means that, as $x \to \infty$,

$$\frac{\lfloor x \rfloor}{x} \to 1$$

but in a rather complicated way, as we can see from its appealing graph in Figure 8.1.

Now we are ready for the proof.

Consider the set of all primes $P = \{p_1, p_2, p_3, \ldots, p_r\}$ less than a positive integer N, then there are

$$\sum_P \left\lfloor \frac{N}{p_1} \right\rfloor$$

of the N numbers that are divisible by at least one of the primes. Similarly, there are

$$\sum_P \left\lfloor \frac{N}{p_1 p_2} \right\rfloor$$

of them divisible by at least two of the primes, etc. Now consider the N^2 pairs of integers, each of which is at most N, $\lfloor N/p_1 \rfloor^2$ of them share p_1 as a divisor, etc., and so

$$\sum_P \left\lfloor \frac{N}{p_1} \right\rfloor^2$$

of them share a single prime as a divisor. Similarly,

$$\sum_P \left\lfloor \frac{N}{p_1 p_2} \right\rfloor^2$$

of them share two primes as divisors, etc. The problem is that in doing this we have to multiply counted numbers: if a number is divisible by three primes, it is divisible by any two or one of them, which is where the inclusion–exclusion theorem comes in. Referring to the letters of its statement, if we write A for the set of pairs sharing a single prime factor, B for the set sharing two prime factors, etc., the inclusion–exclusion principle gives

$$N^2 = \sum_P \left\lfloor \frac{N}{p_1} \right\rfloor^2 - \sum_P \left\lfloor \frac{N}{p_1 p_2} \right\rfloor^2 + \sum_P \left\lfloor \frac{N}{p_1 p_2 p_3} \right\rfloor^2 - \cdots + \Pi_N,$$

where Π_N is the number of co-prime pairs.

Put the other way around

$$\Pi_N = N^2 - \sum_P \left[\frac{N}{p_1} \right]^2 + \sum_P \left[\frac{N}{p_1 p_2} \right]^2 - \sum_P \left[\frac{N}{p_1 p_2 p_3} \right]^2 + \cdots$$

and so

$$\frac{\Pi_N}{N^2} = 1 - \sum_P \left(\frac{1}{N} \left\lfloor \frac{N}{p_1} \right\rfloor \right)^2$$

$$+ \sum_P \left(\frac{1}{N} \left\lfloor \frac{N}{p_1 p_2} \right\rfloor \right)^2 - \sum_P \left(\frac{1}{N} \left\lfloor \frac{N}{p_1 p_2 p_3} \right\rfloor \right)^2 + \cdots .$$

Now we are going to let $N \to \infty$. Using result (3) we have that

$$\lim_{N \to \infty} \frac{1}{N} \left\lfloor \frac{N}{p_1} \right\rfloor = \frac{1}{p_1}$$

and so on for each term. With this and result (2), we have the probability that any two positive integers are co-prime is

$$1 - \sum_P \frac{1}{p_1^2} + \sum_P \frac{1}{p_1^2 p_2^2} - \sum_P \frac{1}{p_1^2 p_2^2 p_3^2} + \cdots$$

$$= 1 - \sum_P \frac{1}{p_1^2} + \sum_P \frac{1}{p_1^2} \frac{1}{p_2^2} - \sum_P \frac{1}{p_1^2} \frac{1}{p_2^2} \frac{1}{p_3^2} + \cdots$$

$$= \left(1 - \frac{1}{p_1^2}\right)\left(1 - \frac{1}{p_2^2}\right)\left(1 - \frac{1}{p_3^2}\right) \cdots$$

$$= \frac{1}{\left(\frac{1}{1 - p_1^{-2}}\right)\left(\frac{1}{1 - p_2^{-2}}\right)\left(\frac{1}{1 - p_3^{-2}}\right) \cdots} = \frac{1}{\zeta(2)},$$

which establishes the result.

Using the Beautiful Formula from p. 60 with $x = 2$ we get

$$\int_0^\infty \frac{u}{e^u - 1} \, du = \zeta(2)\Gamma(2) = \frac{1}{6}\pi^2 \times 1 = \frac{1}{6}\pi^2$$

so, the probability that two integers are co-prime is also

$$1 : \int_0^\infty \frac{u}{e^u - 1} \, du.$$

How is this possible? You may very well ask!

It is also true that the probability that k randomly chosen integers are co-prime is $1 : \zeta(k)$, but that we will not prove!

What Is Gamma ... Exactly?

Constants don't vary—unless they're parameters.

Anon.

9.1 GAMMA EXISTS

We have pretty convincing evidence that the constant γ exists, but no precise proof. Euler did not live in an age of great mathematical rigour and he assuredly was not given to spending his days trying to prove what seemed to him to be intuitively obvious: such thoroughness was to be the stuff of the 19th century. In the 21st, we would be uncomfortable without the security of knowledge that γ really does exist and so we will deal with that matter now.

Given that γ does exist, perhaps the first thing to notice is that we seem to have two definitions of it, one featuring $\ln n$ and the other $\ln(n + 1)$. In fact, they are equivalent and it is more generally the case that

$$\lim_{n \to \infty} \left(\frac{1}{1} + \frac{1}{2} + \frac{1}{3} + \cdots + \frac{1}{n} - \ln(n + \alpha) \right)$$

is independent of $\alpha > -n$, which is easy to see:

$$\lim_{n \to \infty} \left(\frac{1}{1} + \frac{1}{2} + \frac{1}{3} + \cdots + \frac{1}{n} - \ln(n + \alpha) \right)$$

$$= \lim_{n \to \infty} \left(\frac{1}{1} + \frac{1}{2} + \frac{1}{3} + \cdots + \frac{1}{n} - \ln n - \ln(n + \alpha) + \ln n \right)$$

$$= \lim_{n \to \infty} \left(\frac{1}{1} + \frac{1}{2} + \frac{1}{3} + \cdots + \frac{1}{n} - \ln n - \ln \left(1 + \frac{\alpha}{n} \right) \right)$$

$$= \lim_{n \to \infty} \left(\frac{1}{1} + \frac{1}{2} + \frac{1}{3} + \cdots + \frac{1}{n} - \ln n \right).$$

Unsurprisingly, establishing the existence of γ has attracted many proofs and we have chosen one that follows C. W. Barnes of the University of Mississippi.

It is not the shortest but it is elegant, gives an equality for the Euler definition of e, and again makes use of the value of $\zeta(2)$.

We need two very reasonable principles.

(i) For a continuous function $f(x)$,

$$\int_a^b f(x)\,dx = (b-a)f(\xi) \quad \text{for some } \xi \in [a, b].$$

(ii) An increasing sequence of real numbers that is bounded above must approach a limit.

The first of these principles (often known as the first mean value theorem for integration) simply says that the area under the continuous curve $f(x)$ over the interval $[a, b]$ is equal to the area of the rectangle based over the same interval with height determined by some value within that interval, as suggested in Figure 9.1.

The second is a standard (and again reasonable) result of real analysis, relying on the completeness of $\mathbb{R}$.

So, we can start. Using the mean value theorem, we have on the one hand,

$$\int_{1/n+1}^{1/n} \ln x\,dx = \left(\frac{1}{n} - \frac{1}{n+1}\right) \ln c_n = \frac{1}{n(n+1)} \ln c_n, \quad \frac{1}{n+1} < c_n < \frac{1}{n}.$$

On the other hand, if we use the world's most devious integration trick and integrate $\ln x$ by parts we have

$$\int_{1/n+1}^{1/n} \ln x\,dx = \int_{1/n+1}^{1/n} 1 \times \ln x\,dx = [x \ln x - x]_{1/n+1}^{1/n}$$

$$= \left(\frac{1}{n}\ln\frac{1}{n} - \frac{1}{n}\right) - \left(\frac{1}{n+1}\ln\frac{1}{n+1} - \frac{1}{n+1}\right)$$

$$= \left(-\frac{1}{n}\ln n - \frac{1}{n}\right) - \left(-\frac{1}{n+1}\ln(n+1) - \frac{1}{n+1}\right)$$

$$= \frac{1}{n+1}\ln(n+1) - \frac{1}{n}\ln n + \frac{1}{n+1} - \frac{1}{n}$$

$$= \frac{1}{n(n+1)}(n\ln(n+1) - (n+1)\ln n) - \frac{1}{n(n+1)}$$

$$= \frac{1}{n(n+1)}\ln\frac{(n+1)^n}{n^{n+1}} - \frac{1}{n(n+1)}$$

$$= \frac{1}{n(n+1)}\left(\ln\frac{(n+1)^n}{n^{n+1}} - 1\right).$$

Equating the two forms we get

$$\frac{1}{n(n+1)}\ln c_n = \frac{1}{n(n+1)}\left(\ln\frac{(n+1)^n}{n^{n+1}} - 1\right),$$

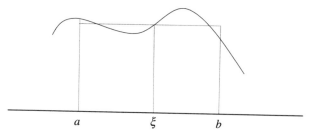

Figure 9.1. The first mean value theorem for integration.

which means that

$$\ln c_n = \ln \frac{(n+1)^n}{n^{n+1}} - 1 \quad \text{or} \quad \ln \frac{(n+1)^n}{n^{n+1}} - \ln c_n = 1$$

and

$$\ln \frac{(n+1)^n / n^{n+1}}{c_n} = 1$$

and this means that

$$\frac{(n+1)^n / n^{n+1}}{c_n} = e.$$

So,

$$e = \frac{(n+1)^n}{n^n} \frac{1}{nc_n} \quad \text{and} \quad e = \left(1 + \frac{1}{n}\right)^n \frac{1}{nc_n}$$

for any positive integer n.

Since

$$\frac{1}{n+1} < c_n < \frac{1}{n}, \qquad n < \frac{1}{c_n} < n+1$$

and so

$$1 < \frac{1}{nc_n} < 1 + \frac{1}{n}$$

and if we write

$$a_n = \frac{1}{nc_n}$$

we have that

$$e = a_n \left(1 + \frac{1}{n}\right)^n, \quad 1 < a_n < 1 + \frac{1}{n} \quad \text{for } n \in \mathbb{N}.$$

This is the equality for e that we mentioned earlier.

If we take the limit,

$$\lim_{n\to\infty} 1 \leqslant \lim_{n\to\infty} a_n \leqslant \lim_{n\to\infty} \left(1 + \frac{1}{n}\right),$$

which makes $\lim_{n\to\infty} a_n = 1$, we recover the Euler definition

$$e = \lim_{n\to\infty} a_n \left(1 + \frac{1}{n}\right)^n = \lim_{n\to\infty} \left(1 + \frac{1}{n}\right)^n.$$

Now change n to r and take logarithms of both sides of

$$e = a_r \left(1 + \frac{1}{r}\right)^r$$

to get

$$1 = \ln a_r + r \ln \left(1 + \frac{1}{r}\right) = \ln a_r + r \ln \left(\frac{r+1}{r}\right)$$

and

$$\frac{1}{r} = \frac{1}{r} \ln a_r + \ln \left(\frac{r+1}{r}\right).$$

Summing gives

$$\sum_{r=1}^{n} \frac{1}{r} = \sum_{r=1}^{n} \frac{1}{r} \ln a_r + \sum_{r=1}^{n} \ln \left(\frac{r+1}{r}\right)$$

$$= \sum_{r=1}^{n} \frac{1}{r} \ln a_r + \sum_{r=1}^{n} (\ln(r+1) - \ln r)$$

$$= \sum_{r=1}^{n} \frac{1}{r} \ln a_r + \ln(n+1)$$

and so

$$\sum_{r=1}^{n} \frac{1}{r} - \ln(n+1) = \sum_{r=1}^{n} \frac{1}{r} \ln a_r.$$

Since each of the $a_r > 1$, the above is an increasing sequence as a function of n; we now show that it is bounded above.

Since $1 < a_n < 1 + 1/n$,

$$\sum_{r=1}^{n} \frac{1}{r} - \ln(n+1) = \sum_{r=1}^{n} \frac{1}{r} \ln a_r < \sum_{r=1}^{n} \frac{1}{r} \ln \left(1 + \frac{1}{r}\right).$$

It is geometrically clear that $\ln(1 + x) < x$ for $x > 0$ and so

$$\ln \left(1 + \frac{1}{r}\right) < \frac{1}{r}$$

and

$$\sum_{r=1}^{n} \frac{1}{r} - \ln(n+1) < \sum_{r=1}^{n} \frac{1}{r^2} < \frac{\pi^2}{6}$$

and we have the promised re-emergence of $\zeta(2)$.

So, the left-hand side is bounded above as well as increasing and it therefore tends to a limit: recalling the earlier observation that

$$\gamma = \lim_{n \to \infty} \left(\frac{1}{1} + \frac{1}{2} + \frac{1}{3} + \cdots + \frac{1}{n} - \ln(n + \alpha) \right),$$

put $\alpha = 1$ and we are finished.

9.2 GAMMA IS... WHAT NUMBER?

Now we know that γ exists, it is not unreasonable to ask for its value and since its exact nature remains one of its mysteries, we are bound to concentrate on approximations. We have already seen on pp. 47 and 48 that it lies between 0 and 1 and looks likely to be around 0.5.

To find a decimal expression for γ we could simply evaluate $\gamma_n = H_n - \ln n$ for increasing values of n, but the convergence is extremely slow; for example, $\gamma_{100} = 0.582\,207\,331\,651\,53\ldots$, which is accurate only to one decimal place, and $\gamma_{1\,000\,000} = 0.577\,216\,164\,901\,481\ldots$ is accurate only to five decimal places. With each component equally reluctantly diverging to infinity, it seems a shame that they combine to an equally reluctant convergence. The reason is exposed by the inequality

$$\frac{1}{2(n + 1)} < \gamma_n - \gamma < \frac{1}{2n}, \quad n \in \mathbb{N}.$$

Assuming that this is true, if we want an accuracy of m decimal places, we require $\gamma_n - \gamma < 5 \times 10^{-m-1}$ and so

$$\frac{1}{2n} < 5 \times 10^{-m-1},$$

which means that $n > 10^m$, and the strict inequality is needed, since

$$\gamma_n - \gamma > \frac{1}{2(n + 1)} = \frac{1}{2n} \left(1 + \frac{1}{n} \right)^{-1} > \frac{1}{2n} \left(1 - \frac{1}{n} \right)$$

and if $n = 10^m$,

$$\gamma_n - \gamma > \frac{1}{2 \times 10^m} \left(1 - \frac{1}{10^m} \right) = \frac{5}{10^{m+1}} \left(1 - \frac{1}{10^m} \right)$$

$$= \frac{5}{10^{m+1}} \left(\frac{10^m - 1}{10^m} \right) = 4.\underbrace{999\,999\,999\,99}_{(m-1)\ \text{times}} 5 \times 10^{-(m+1)},$$

which guarantees that the approximation is incorrect in the mth decimal place.

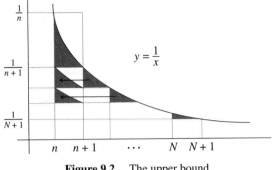

Figure 9.2. The upper bound.

Having used the inequality, we give R. M. Young's proof of it, which uses the technique we adopted on p. 43 to describe the behaviour of the Zeta function as $x \to 1^+$. Referring to Figure 9.2,

$$\sum_{n}^{N} \text{shaded areas touching the curve}$$

$$= \left(\int_{n}^{n+1} \frac{1}{x}\,dx - \frac{1}{n+1} \right) + \left(\int_{n+1}^{n+2} \frac{1}{x}\,dx - \frac{1}{n+2} \right)$$

$$+ \cdots + \left(\int_{N-1}^{N} \frac{1}{x}\,dx - \frac{1}{N} \right)$$

$$= \int_{n}^{N} \frac{1}{x}\,dx - \sum_{r=1}^{N-n} \frac{1}{n+r} = \int_{n}^{N} \frac{1}{x}\,dx - \left(\sum_{r=1}^{N} \frac{1}{r} - \sum_{r=1}^{n} \frac{1}{r} \right)$$

$$= \left(\ln N - \sum_{r=1}^{N} \frac{1}{r} \right) - \left(\ln n - \sum_{r=1}^{n} \frac{1}{r} \right).$$

Now let $N \to \infty$ and we have, by definition,

$$\sum_{n}^{\infty} \text{shaded areas} = -\gamma + \gamma_n = \gamma_n - \gamma.$$

If we now horizontally translate the shaded regions so that they all lie in the first rectangle between n and $n + 1$, we see that each region has an area less than one-half of the rectangle enclosing it (owing to the concavity of $1/x$) and so the total area of all of the regions is less than one-half of the area of the first rectangle, which is clearly $1/n$, which means that

$$\gamma_n - \gamma < \frac{1}{2n}.$$

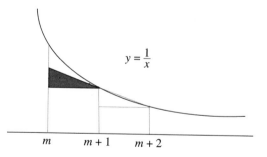

Figure 9.3. The lower bound.

To achieve the lower bound, embed a right-angled triangle in each region as shown in Figure 9.3, where the hypotenuse is the continuation of the hypotenuse of the circumscribed triangle to the right. The shaded triangle and the circumscribed one used to define it are clearly congruent and since the area of the latter is

$$\frac{1}{2}\left(\frac{1}{m+1} - \frac{1}{m+2}\right),$$

summing these gives

$$\gamma_n - \gamma = \sum_{n}^{\infty} \text{shaded areas} > \frac{1}{2}\sum_{m=n}^{\infty}\left(\frac{1}{m+1} - \frac{1}{m+2}\right) = \frac{1}{2(n+1)}.$$

And so we have

$$\frac{1}{2(n+1)} < \gamma_n - \gamma < \frac{1}{2n},$$

as required.

9.3 A SURPRISINGLY GOOD IMPROVEMENT

The above bound relates to the $\ln n$ form of the definition of γ and even though in the limit we have seen that

$$\gamma = \lim_{n \to \infty}\left(\frac{1}{1} + \frac{1}{2} + \frac{1}{3} + \cdots + \frac{1}{n} - \ln(n+\alpha)\right) \quad \text{for any } \alpha > -n,$$

we might expect the choice of α to influence the approximations for finite values of n, and so it does, as we can see if we construct an error function $\varepsilon_n(\alpha)$, defined by

$$\varepsilon_n(\alpha) = \frac{1}{1} + \frac{1}{2} + \frac{1}{3} + \cdots + \frac{1}{n} - \ln(n+\alpha) - \gamma, \quad n \geqslant 1, \quad \alpha > -n,$$

where γ can be represented in decimal form to any degree of accuracy using its original definition (given we have the patience and calculating accuracy

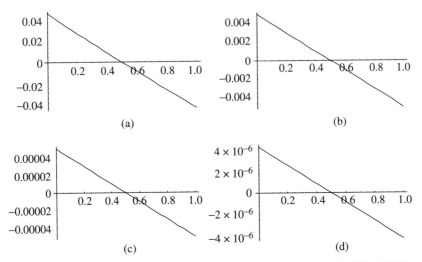

Figure 9.4. The error function near its zero. (a) $n = 10$, zero is $0.503\,962\,732\,569\,747$; (b) $n = 100$, zero is $0.500\,414\,587\,370\,329$; (c) $n = 10\,000$, zero is $0.500\,004\,166\,069\,63$; (d) $n = 100\,000$, zero is $0.500\,000\,401\,909\,347$.

needed). Of course, for all α, $\varepsilon_n(\alpha) \to 0$ as $n \to \infty$, but it is interesting (and surprising) to look at the function for fixed n as α varies.

If we differentiate with respect to α, we get

$$\frac{d\varepsilon_n(\alpha)}{d\alpha} = -\frac{1}{n + \alpha}$$

and so the function will forever (but diminishingly) decrease from $+\infty$ at its vertical asymptote at $\alpha = -n$, to $-\infty$ as α increases, making its zero unique.

Figure 9.4 concentrates on the interval $0 \leqslant \alpha \leqslant 1$ and over this small interval inevitably give a false impression of linearity but the eye is drawn to the zero at a value of α ever closer to 0.5.

If we take the strong hint provided by these plots, we would reasonably take $\alpha = \frac{1}{2}$ if we wish to minimize the error for any n and so consider the form of the definition as

$$\gamma = \lim_{n \to \infty} \left(\frac{1}{1} + \frac{1}{2} + \frac{1}{3} + \cdots + \frac{1}{n} - \ln\left(n + \frac{1}{2}\right) \right) = \lim_{n \to \infty} \rho_n.$$

Recall that, with $\alpha = 0$, γ_{100} is accurate only to one decimal place and $\gamma_{1\,000\,000}$ only to five decimal places; now, with $\alpha = \frac{1}{2}$, $\rho_{100} = 0.577\,219\,790\,140\,49$ and $\rho_{1\,000\,000} = 0.577\,215\,664\,900\,631$, and these are accurate to five and eleven decimal places, respectively.

The explanation for this huge improvement is that

$$\frac{1}{24(n + 1)^2} < \rho_n < \frac{1}{24n^2}$$

and we give Duane W. DeTemple's proof of the result:

$$\rho_n - \rho_{n+1} = H_n - \ln(n + \tfrac{1}{2}) - H_{n+1} + \ln(n + \tfrac{3}{2})$$

$$= -\frac{1}{n+1} - \ln(n + \tfrac{1}{2}) + \ln(n + \tfrac{3}{2}).$$

Define the function

$$f(x) = -\frac{1}{x+1} - \ln(x + \tfrac{1}{2}) + \ln(x + \tfrac{3}{2}), \quad x > 0,$$

$$f'(x) = \frac{1}{(1+x)^2} - \frac{1}{(x + \tfrac{1}{2})} + \frac{1}{(x + \tfrac{3}{2})}$$

$$= \frac{1}{(1+x)^2} - \frac{1}{(x + \tfrac{1}{2})(x + \tfrac{3}{2})} = \frac{x^2 + 2x + \tfrac{3}{4} - (1 + 2x + x^2)}{(1+x)^2(x + \tfrac{1}{2})(x + \tfrac{3}{2})}$$

$$= -\tfrac{1}{4}(x + 1)^{-2}(x + \tfrac{1}{2})^{-1}(x + \tfrac{3}{2})^{-1}$$

and since $(x + \tfrac{3}{2})^{-1} < (x + 1)^{-1} < (x + \tfrac{1}{2})^{-1}$, $-f'(x) < \tfrac{1}{4}(x + \tfrac{1}{2})^{-4}$.
As $f(\infty) = 0$,

$$f(k) = -\int_k^\infty f'(x)\,dx < \frac{1}{4}\int_k^\infty (x + \tfrac{1}{2})^{-4}\,dx$$

$$= -\tfrac{1}{12}[(x + \tfrac{1}{2})^{-3}]_k^\infty = \tfrac{1}{12}(k + \tfrac{1}{2})^{-3}.$$

Since $(k + \tfrac{1}{2})^2 > k(k+1)$, $(k + \tfrac{1}{2})^4 > k^2(k+1)^2$ and $(k + \tfrac{1}{2})^{-4} < 1/(k^2(k+1)^2)$
and so

$$(k + \tfrac{1}{2})^{-3} < \frac{1}{k^2(k+1)^2}(k + \tfrac{1}{2})$$

$$= \frac{1}{2}\frac{2k+1}{k^2(k+1)^2} = \frac{1}{2}\left(\frac{1}{k^2} - \frac{1}{(k+1)^2}\right)$$

$$= \int_k^{k+1} x^{-3}\,dx,$$

$$\therefore \quad \rho_n - \gamma = \sum_{k=n}^\infty (\rho_k - \rho_{k+1})$$

$$= \sum_{k=n}^\infty f(k) < \frac{1}{12}\sum_{k=n}^\infty (k + \tfrac{1}{2})^{-3} < \frac{1}{12}\int_n^\infty x^{-3}\,dx$$

$$= \frac{1}{24n^2}.$$

And we have the inequality one way around.
The other half is found in the following way.

Since $(x + \frac{1}{2})(x + \frac{3}{2}) = x^2 + 2x + \frac{3}{4} < (x + 1)^2$, $(x + \frac{1}{2})^{-1}(x + \frac{3}{2})^{-1} > (x + 1)^{-2}$ and so $-f'(x) > \frac{1}{4}(x + 1)^{-4}$.

As before,

$$f(k) = -\int_k^\infty f'(x)\,dx > \frac{1}{4}\int_k^\infty (x + 1)^{-4}\,dx$$
$$= -\frac{1}{12}[(x + 1)^{-3}]_k^\infty = \frac{1}{12}(k + 1)^{-3}.$$

So,

$$\rho_n - \gamma = \sum_{k=n}^\infty (\rho_k - \rho_{k+1})$$

$$= \sum_{k=n}^\infty f(k) > \frac{1}{12}\sum_{k=n}^\infty (k + 1)^{-3} > \frac{1}{12}\int_{n+1}^\infty x^{-3}\,dx$$

$$= \frac{1}{24(n + 1)^2}.$$

And we are done.

Again, if we wanted an accuracy of m decimal places, we require $\rho_n - \gamma < 5 \times 10^{-m-1}$, and so

$$\frac{1}{24n^2} < 5 \times 10^{-m-1} \quad \text{and} \quad n > \sqrt{\frac{10^{m+1}}{5 \times 24}} \approx 0.288\,675 \times 10^{m/2}.$$

Again, the strict inequality is needed, since

$$\rho_n - \gamma > \frac{1}{24(n + 1)^2} = \frac{1}{24n^2}\left(1 + \frac{1}{n}\right)^{-2} > \frac{1}{24n^2}\left(1 - \frac{2}{n}\right)$$

and so, if $n = \sqrt{10^{m+1}/(5 \times 24)}$,

$$\delta_n - \gamma > \frac{5 \times 24}{24 \times 10^{m+1}}\left(1 - \frac{2\sqrt{120}}{10^{(m+1)/2}}\right) = 4.\underbrace{999\,999\,999}_{(m-1)\text{ times}}45 \cdots \times 10^{-(m+1)},$$

which again guarantees that the approximation is incorrect in the mth decimal place.

9.4 THE GERM OF A GREAT IDEA

Stretching the properties of $\approx$ perhaps a little too much, we can rewrite the statement

$$\gamma_n = H_n - \ln n \approx \frac{1}{2} + \frac{1}{2n} \quad \Rightarrow \quad \gamma \approx 0.5$$

as

$$\gamma_n - \gamma \approx \frac{1}{2n} \quad \text{or} \quad \gamma \approx \gamma_n - \frac{1}{2n},$$

which for $n = 1000$ gives $\gamma \approx 0.577\,215\,581\,568\,204\ldots$, which is accurate to six decimal places, and for $n = 1\,000\,000$ gives $\gamma \approx 0.577\,215\,664\,901\,481\ldots$, which is accurate to twelve decimal places; it may not be rigorous but we are on the right track in approximating γ! Actually, this is the first of a series of approximations, which continue mysteriously as

$$\gamma \approx \gamma_n - \frac{1}{2n} + \frac{1}{12n^2} - \frac{1}{120n^4} + \frac{1}{252n^6} - \frac{1}{240n^8} + \cdots + \frac{1}{12n^{14}} \cdots$$

the mystery deepening with the knowledge that the term involving n^{12} has -691 on the top and $32\,760$ on the bottom.

In fact, the approximation may be written more fully as

$$\gamma \approx \gamma_n - \frac{1}{2n} + \sum_{r=1}^{\infty} \frac{B_{2r}}{2r} \frac{1}{n^{2r}},$$

which is a special case of the Euler–Maclaurin summation formula, where B_{2r} are known as the Bernoulli Numbers—both of which we will look at next.

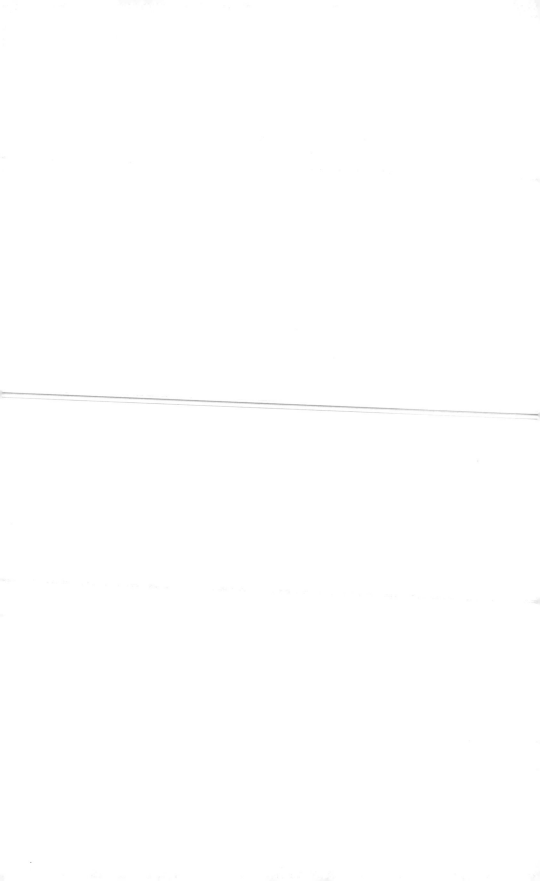

Gamma as a Decimal

A mathematician is a blind man in a dark room looking for a black cat which isn't there.

Charles Darwin (1809–1882)

10.1 BERNOULLI NUMBERS

Our earlier focus on the Zeta series has meant that, in terms of the summation of series, we have in a way started on the second rung of the ladder, with the first occupied by the family $1^k + 2^k + 3^k + \cdots + n^k$ for $k \in \mathbb{N}$. In 1784, at the age of seven, Gauss had famously summed the integers from 1 to 100 in seconds (to the amazement of his teacher) when he noticed that the series could be thought of as 50 pairs of numbers each summing to 101; of course, the young genius could not have known that the ancient Greeks, Hindus and Arabs each had rules which amounted to the sum for $k \leqslant 4$, nor would he have been aware of the work of Johann Faulhaber (1580–1635). Known in his time as 'The Great Arithmetician (or weaver) of Ulm', Faulhaber was indeed trained as a weaver but his mathematical prowess brought his appointment as the city's mathematician and surveyor, who designed waterwheels, fortifications and surveying instruments and who associated and collaborated with the likes of Kepler, Descartes and Napier; he also prepared the first German publication of Briggs's logarithms. In fact, he was a 'Cossist' more than an 'Arithmetician', whose 1631 publication *Academiae Algebrae* contained not only the sums up to $k = 17$ but also the important observation that

$$
1^k + 2^k + 3^k + \cdots + n^k =
\begin{cases}
\text{a polynomial in } n(n+1) & k \text{ odd,} \\
(2n+1) \times \text{a polynomial in } n(n+1) & k \text{ even.}
\end{cases}
$$

(The term *Cossist* derives from the Italian word 'cosa', meaning 'thing'; the mathematicians of the time used the word to represent an unknown quantity, we would use the word 'algebraist'.) In 1636 Fermat had need of an answer as he calculated such sums in his development of the *quadrature* of $f(x) = x^k$, prior to Newton's calculus. He found a recurrence relation relating the sum for k with the sums for $k - 1, k - 2, \ldots$, which was ingenious but soon intractable and, although improved on in 1654 by Blaise Pascal (1623–1662), the problem had to wait until the next century to be solved by one of its greatest mathematical names.

The first few expressions can be written:

$$1 + 2 + 3 + \cdots + n = \tfrac{1}{2}n(n + 1),$$
$$1^2 + 2^2 + 3^2 + \cdots + n^2 = (2n + 1)\tfrac{1}{6}n(n + 1),$$
$$1^3 + 2^3 + 3^3 + \cdots + n^3 = \tfrac{1}{4}[n(n + 1)]^2,$$
$$1^4 + 2^4 + 3^4 + \cdots + n^4 = (2n + 1)\tfrac{1}{30}n(n + 1)[3n(n + 1) - 1],$$

which reveals nothing more than Faulhaber's observation and the very pretty relationship

$$1^3 + 2^3 + 3^3 + \cdots + n^3 = (1 + 2 + 3 + \cdots + n)^2.$$

It was Jacob Bernoulli who solved the problem and the solution was announced to the world in his famous treatise *Ars Conjectandi*, posthumously published in 1713. In listing the results to $k = 10$, Bernoulli described the pattern that mattered; somewhat generously inferring that others might also have the same powers of insight, he wrote (without proof):

> Whoever will examine the series as to their regularity may be able to continue the table. Taking c to be the power of any exponent or

$$\int n^c \infty \frac{1}{c + 1}n^{c+1} + \tfrac{1}{2}n^c + \tfrac{1}{2}cAn^{c-1}$$
$$+ \frac{c.c - 1.c - 2}{2.3.4}Bn^{c-3} + \frac{c.c - 1.c - 2.c - 3.c - 4}{2.3.4.5.6}Cn^{c-5}$$
$$+ \frac{c.c - 1.c - 2.c - 3.c - 4.c - 5.c - 6}{2.3.4.5.6.7.8}Dn^{c-7},$$

> and so on, the exponents of continually decreasing by 2 until n or nn is reached. The capital letters A, B, C, D denote in order the coefficients of the last terms in the expressions for $\int nn$, $\int n^4$, $\int n^6$, $\int n^8$, etc., namely, A is equal to $1/6$, B is equal to $-1/30$, C is equal to $1/42$, D is equal to $-1/30$.

> These coefficients are such that each one completes the others in the same expression to unity. Thus D must have the value $-1/30$ because

$$\tfrac{1}{9} + \tfrac{1}{2} + \tfrac{2}{3} - \tfrac{7}{15} + \tfrac{2}{9} + (+D) - \tfrac{1}{30} = 1.$$

With the help of this table it took me less than half of a quarter of an hour to find that the tenth powers of the first 1000 numbers being added together will yield the sum

$$91\,409\,924\,241\,424\,243\,424\,241\,924\,242\,500.$$

From this it will become clear how useless was the work of Ismael Bullialdus spent on the compilation of his voluminous *Arithmatica Infinitorum* in which he did nothing more than compute with immense labour the sums of the first six powers, which is only a part of what we have accomplished in the space of a single page.

The withering comment regarded the prodigious efforts of Ismael Bullialdus (1605–1694), who needed a six-volume opus to achieve the result for the first six powers. Notice the use of the 'backwards proportional sign' for '=', of nn for n^2, of what is now the integral sign for summation (Euler's influence had yet to take affect), of a dot for multiplication and the implied brackets in the expressions involving c. Incidentally, he erroneously gave the coefficient of n^2 for $k = 9$ as $-\frac{1}{12}$ *rather* than its correct value of $-\frac{3}{20}$. In identifying what he called A, B, C, D, ... Bernoulli had isolated the numbers in the expansion which are independent of the power and if we begin to list them all, including those which are zero, we have the appropriately named (by Euler) Bernoulli Numbers B_0, B_1, B_2, ...

$$1,\quad \frac{1}{2},\quad \frac{1}{6},\quad 0,\quad -\frac{1}{30},\quad 0,\quad \frac{1}{42},\quad 0,\quad -\frac{1}{30},\quad 0,\quad \frac{5}{66},\quad 0,\quad \dots$$

with a pattern anything but transparent. The next term is 691/2730 and, to emphasize the point, the sequence continues

$$\frac{7}{6},\quad -\frac{3617}{510},\quad \frac{43\,867}{798},\quad \dots$$

In more modern guise, and using the standard notation

$$\binom{n}{r} = \frac{n!}{r!(n-r)!}$$

for the binomial coefficients, Bernoulli was really saying that

$$1 + 2 + 3 + \cdots + n = \tfrac{1}{2}n^2 + \tfrac{1}{2}n = \frac{1}{2}\left(\binom{2}{0}B_0 n^2 + \binom{2}{1}B_1 n\right),$$

$$1^2 + 2^2 + 3^2 + \cdots + n^2 = \tfrac{1}{3}n^3 + \tfrac{1}{2}n^2 + \tfrac{1}{6}n$$

$$= \frac{1}{3}\left(\binom{3}{0}B_0 n^3 + \binom{3}{1}B_1 n^2 + \binom{3}{2}B_2 n\right),$$

$$1^3 + 2^3 + 3^3 + \cdots + n^3 = \tfrac{1}{4}n^4 + \tfrac{1}{2}n^3 + \tfrac{1}{4}n^2$$

$$= \frac{1}{4}\left(\binom{4}{0}B_0 n^4 + \binom{4}{1}B_1 n^3\right.$$

$$\left. + \binom{4}{2}B_2 n^2 + \binom{4}{3}B_3 n\right),$$

$$1^4 + 2^4 + 3^4 + \cdots + n^4 = \tfrac{1}{5}n^5 + \tfrac{1}{2}n^4 + \tfrac{1}{3}n^3 - \tfrac{1}{30}n$$

$$= \frac{1}{5}\left(\binom{5}{0}B_0 n^5 + \binom{5}{1}B_1 n^4 + \binom{5}{2}B_2 n^3\right.$$

$$\left. + \binom{5}{3}B_3 n^2 + \binom{5}{4}B_4 n\right).$$

Although the Bernoulli Numbers lack an obvious pattern, they do possess a recursive definition, which Bernoulli announced through his computation of D. His explanation related to the expansion for $k = 8$, which he gave as

$$1^8 + 2^8 + 3^8 + \cdots + n^8 = \frac{1}{9}n^9 + \frac{1}{2}n^8 + \frac{2}{3}n^7 - \frac{7}{15}n^5 + \frac{2}{9}n^3 - \frac{1}{30}n.$$

Noting that for $n = 1$ both sides must be 1, it is possible to solve for any one of the numbers in terms of the others and this he did for his D, using for us a slightly strange algebraic form. Every odd-numbered Bernoulli Number (other than the first) is 0 and of course every even one can be found from the recurrence relation, albeit tediously. There are plenty of alternative ways of generating them and they appear as part of the coefficients of any number of expansions, for example

$$\frac{x}{e^x - 1}$$

(given by Euler), and they can be efficiently generated in terms of what are known as 'tangent numbers' but no one would describe them as cooperative. Euler computed them up to B_{30}, in 1840 Ohm extended this to B_{62} and the following year Adams computed them to B_{124}—the numerator of which has 110 digits (contrasting with the denominator, which is simply the number 30). The calculations cry out for the computational aids that we now take for granted, an application of computers that was presaged in 1843 by Augusta Ada King, Countess Lovelace (and daughter of Lord Byron), who suggested to Charles Babbage that he produce a 'plan' for their calculation, using his Analytical Engine. Later, in her annotated translation of a publication of one Luigi Federico Menabrea (one time Professor of Mechanics at Turin and later the Italian premier) dealing with ideas relating to the Analytical Engine, she described several such 'plans', which might be considered to be the earliest recorded computer programs for a device which she romantically posited 'weaves algebraic patterns, just as the Jacquard-loom weaves flowers and leaves'.

With Fermat's Last Theorem finally laid to rest, it is of no more (and no less) than historic interest that the Bernoulli Numbers have played their role in its attempted resolution. In 1850 Ernst Kummer (1810–1893) proved the theorem for all powers which were 'regular' primes, with the definition of 'regular' the elegant 'a prime p is regular if and only if it does not divide the numerator of $B_2, B_4, B_6, \ldots, B_{p-3}$'. It is known that the number of irregular primes is infinite but unfortunately whether the same is true for regular primes is unknown (the first irregular prime is 37 since $B_{32} = -208\,360\,028\,141 \times 37/510$).

10.2 EULER–MACLAURIN SUMMATION

We have noted that

$$\gamma = \lim_{n \to \infty} \left(\frac{1}{1} + \frac{1}{2} + \frac{1}{3} + \cdots + \frac{1}{n} - \ln n \right)$$

can be thought of as the difference between the sum and the integral of the function $f(x) = 1/x$, in that

$$\gamma = \lim_{n \to \infty} \left(\frac{1}{1} + \frac{1}{2} + \frac{1}{3} + \cdots + \frac{1}{n} - \ln n \right)$$

$$= \lim_{n \to \infty} \left(\sum_{k=1}^{n} \frac{1}{k} - \int_{1}^{n} \frac{1}{x} \, dx \right)$$

$$= \lim_{n \to \infty} \left(\sum_{k=1}^{n} f(k) - \int_{1}^{n} f(x) \, dx \right)$$

and if we relegate γ to secondary importance we could write

$$\sum_{k=1}^{n} \frac{1}{k} - \int_{1}^{n} \frac{1}{x} \, dx \approx \gamma$$

and so

$$\sum_{k=1}^{n} \frac{1}{k} \approx \int_{1}^{n} \frac{1}{x} \, dx + \gamma.$$

With this emphasis we are approximating a sum by an integral and even though integration can be tough it can also be significantly easier than summation: we may be on to a good idea here. We are, but in developing the initiative Euler and Colin Maclaurin (1698–1746) have beaten us by the best part of 300 years, producing what has become known as the Euler–Maclaurin summation formula. We will not prove it but we will use it for our purposes, and it has wide application in many areas of mathematics, perhaps most all in numerical analysis, analytic number theory and the general theory of asymptotic expansions. In 1736 Euler

85

had developed both the simplest form of the formula and, later in the year, the general form, quite independently of Maclaurin, who had published it in his *Treatise of Fluxions* of 1742. In one of its general forms it states

$$\sum_{k=1}^{n} f(k) = \int_{1}^{n} f(x)\, dx + \tfrac{1}{2}(f(1) + f(n))$$

$$+ \sum_{k=1}^{m} \frac{B_{2k}}{(2k)!}(f^{2k-1}(n) - f^{2k-1}(1)) + R_n(f, m),$$

$$R_n(f, m) \leqslant \frac{2}{(2\pi)^{2m}} \int_{1}^{n} |f^{2m+1}(x)|\, dx,$$

where the B_{2k} are, of course, the Bernoulli Numbers and the $(2k-1)$th 'powers' of the function are in fact the $(2k-1)$th derivatives of it. Use of the expansion can be subtle and here is a case when neglect of the remainder term $(R_n(f, m))$ can be perilous since for most functions that appear in applications the series diverges; fortunately, it is usual that not many terms are needed to achieve good accuracy and so the approximations provided by the series are generally excellent. This fact troubled Euler and it was left to Siméon Poisson (1781–1840) in 1823 to pay serious attention to the remainder term.

10.3 Two Examples

1.   As a first move, we can gain some confidence by showing that the Euler–Maclaurin formula gives the result we would expect for $f(x) = x^3$. The derivatives are, of course, $f'(x) = 3x^2$, $f''(x) = 6x$ and $f'''(x) = 6$; the remaining derivatives are zero and so the error term is zero too:

$$\sum_{k=1}^{n} k^3 = \int_{1}^{n} x^3\, dx + \tfrac{1}{2}(1^3 + n^3) + \frac{B_2}{2!}(3n^2 - 3 \times 1^2) + \frac{B_4}{4!}(6 - 6)$$

$$= \tfrac{1}{4}n^4 - \tfrac{1}{4} + \tfrac{1}{2} + \tfrac{1}{2}n^3 + \tfrac{1}{2} \times \tfrac{1}{6}(3n^2 - 3)$$

$$= \tfrac{1}{4}n^4 + \tfrac{1}{2}n^3 + \tfrac{1}{4}n^2 = (\tfrac{1}{2}n(n + 1))^2.$$

2. As a second application of the formula, we will look at a justly famous (if misnamed) result for approximating $n!$ for large n. This time take $f(x) = \ln x$ to get

$$f'(x) = \frac{1}{x}, \quad f''(x) = -\frac{1}{x^2},$$

$$f'''(x) = \frac{2}{x^3}, \quad \cdots, \quad f^{(n)}(x) = (-1)^{n-1}\frac{(n-1)!}{x^n}.$$

This time we will suppress the error term to get

$$\sum_{k=1}^{n} \ln k = \int_{1}^{n} \ln x \, dx + \tfrac{1}{2}(\ln 1 + \ln n)$$

$$+ \frac{B_2}{2!}\left(\frac{1}{n} - \frac{1}{1}\right) + \frac{B_4}{4!}\left(\frac{2}{n^3} - \frac{2}{1^3}\right) + \frac{B_6}{6!}\left(\frac{24}{n^5} - \frac{24}{1^5}\right) + \cdots.$$

Using standard properties of logarithms on the left-hand side and that meanest integration-by-parts trick, on the right-hand side (writing $\ln x = 1 \times \ln x$) we get

$$\ln n! = n \ln n - n + \frac{1}{2}\ln n + \frac{1}{12n} - \frac{1}{360n^3} + \frac{1}{1260n^5} + C_n + \cdots,$$

where C_n is the constant to this number of terms. Now exponentiate both sides to get

$$n! = n^n e^{-n} \sqrt{n} e^{C_n} \exp\left(\frac{1}{12n} - \frac{1}{360n^3} + \frac{1}{1260n^5} + \cdots\right).$$

Using the Taylor expansion of e^x then gives (which the reader can check!)

$$n! = n^n e^{-n} \sqrt{n} e^{C_n} \left(1 + \frac{1}{12n} + \frac{1}{288n^2} - \frac{139}{51840n^3}\right.$$
$$\left. - \frac{571}{2\,488\,320n^4} + \frac{163\,879}{209\,018\,880n^5} + \cdots\right),$$

which could be an excellent approximation to $n!$, if only we knew the asymptotic value of e^{C_n}, given that the limit exists. The series is the well-known 'Stirling approximation', which James Stirling (1692–1770) published to the first eight terms in his most important work *Methodus Differentialis* of 1730. In fact, his interest was in the logarithms of factorials and he left the series in its logarithmic form, computing $\log_{10} 1000!$ to 10 decimal places, using an approximation for the constant. In the same year Abraham de Moivre (1667–1754) published *Miscellanea Analytica*, which, apart from anything else, contained his own (later to be corrected) table of logarithms, his own form of the approximation and a proof of the constant's existence. It would be some years before Stirling would be able to find the constant in exact form and in doing so found it to be $e^{C_n} \xrightarrow[n\to\infty]{} \sqrt{2\pi}$ and the series is then

$$n! = n^n e^{-n} \sqrt{2\pi n}\left(1 + \frac{1}{12n} + \frac{1}{288n^2} - \frac{139}{51840n^3}\right.$$
$$\left. - \frac{571}{2\,488\,320n^4} + \frac{163\,879}{209\,018\,880n^5} + \cdots\right).$$

Here is a case when the error term does misbehave, since for any fixed n it decreases as we take more terms to a point when it starts to increase; fortunately, with m fixed, as n increases the error term does tend to zero and we obtain ever better approximations to $n!$.

We will need Stirling's approximation several times and while it is handy, we can sensibly mention another constant that in a way arises from it and which we will also mention again later. We can rewrite the first-order approximation as

$$\frac{n!}{n^{n+1/2}e^{-n}} \approx \sqrt{2\pi},$$

meaning that

$$\lim_{n\to\infty} \frac{n!}{n^{n+1/2}e^{-n}} = \sqrt{2\pi}.$$

Replacing $n!$ by some other asymptotically large quantity and dividing by an appropriate expression can lead to a constant other than $\sqrt{2\pi}$. In particular, the nice $0^0 1^1 2^2 3^3 \cdots n^n$ and $f(n) = n^{n^2/2+n/2+1/12} e^{-n^2/4}$ combine so that

$$\lim_{n\to\infty} \frac{0^0 1^1 2^2 3^3 \cdots n^n}{f(n)} = A,$$

the Glaisher–Kinkelin constant, which is about $1.282\,427\,13\ldots$.

Exotic it may be, but useful it is too—as we will see!

10.4 THE IMPLICATIONS FOR GAMMA

If we apply the Euler–Maclaurin formula to $f(x) = 1/x$, we get

$$f'(x) = -\frac{1}{x^2}, \quad f''(x) = \frac{2}{x^3},$$

$$f'''(x) = -\frac{3 \times 2}{x^4}, \quad \cdots, \quad f^{(r)}(x) = (-1)^r \frac{r!}{x^{r+1}}.$$

This means that

$$\sum_{k=1}^{n} \frac{1}{k} = \ln n + \frac{1}{2}\left(\frac{1}{1} + \frac{1}{n}\right)$$

$$+ \sum_{k=1}^{m} \frac{B_{2k}}{(2k)!}\left((-1)^{2k-1}\frac{(2k-1)!}{n^{2k}} - (-1)^{2k-1}(2k-1)!\right)$$

$$+ R_n(f, m)$$

$$= \frac{1}{2}\left(\frac{1}{1} + \frac{1}{n}\right) + \sum_{k=1}^{m} \frac{B_{2k}}{2k}\left(1 - \frac{1}{n^{2k}}\right) + R_n(f, m),$$

with the factorial cancelling and the odd power of -1 replaced by -1 itself.

But

$$\gamma = \lim_{n \to \infty} \left(\sum_{k=1}^{n} \frac{1}{k} - \ln n \right)$$

$$= \frac{1}{2} + \sum_{k=1}^{m} \frac{B_{2k}}{2k} + R_\infty(f, m)$$

and so

$$\sum_{k=1}^{n} \frac{1}{k} = \ln n + \gamma + \frac{1}{2n} - \sum_{k=1}^{m} \frac{B_{2k}}{2k} \frac{1}{n^{2k}} + (R_n(f, m) - R_\infty(f, m))$$

and looking at the first few terms (and ignoring the error term) we have

$$\sum_{k=1}^{n} \frac{1}{k} = \ln n + \gamma + \frac{1}{2n} - \frac{1}{12n^2} + \frac{1}{120n^4} - \frac{1}{252n^6} + \cdots$$

and we have

$$\gamma = \sum_{k=1}^{n} \frac{1}{k} - \ln n - \frac{1}{2n} + \frac{1}{12n^2} - \frac{1}{120n^4} + \frac{1}{252n^6} + \cdots .$$

And here is the generalization of the series for γ that has been suggested on p. 79. Euler used the series up to the term $1/12n^{14}$ and with $n = 10$, $H_{10} = 2.928\,968\,253\,968\,253\,9$ and $\ln 10 = 2.302\,585\,092\,994\,045\,684$ to compute γ to those 16 decimal places $0.577\,215\,664\,901\,532\,5 \ldots$.

Of course, the desire to extend the accuracy of the estimate was great and, in 1790, the Italian geometer Lorenzo Mascheroni (1750–1800) published in *Adnotationes ad calculum integrale Euleri* an approximation of γ to 32 decimal places, which he had calculated in a similar way; the estimate then became $0.577\,215\,664\,901\,532\,860\,618\,1 \ldots$. This was all well and good until 1809, when Johann von Soldner (1766–1833) used his

$$Li(x) = \int_2^x \frac{1}{\ln x} \, dx$$

function (which will engage our attention later) to give the value

$$0.577\,215\,664\,901\,532\,860\,606\,5 \ldots ,$$

which differs in that underlined 20th decimal place (and after). The matter was resolved (but the confusion not removed) when, in 1812, the inimitable Gauss prevailed on the 19-year-old calculating prodigy F. G. B. Nicolai (1793–1846) to check the results. This he did, using the Euler–Maclaurin summation formula with $n = 50$ and recalculating with $n = 100$ to evaluate γ to 40 decimal places—and finding agreement with Soldner. In spite of this, both values were in circulation (and even appeared together in one publication), which led subsequent indefatigable calculators (again using Euler–Maclaurin summation) independently to provide their own confirmation of Soldner's estimate. Mascheroni's permanent contribution to γ's story (apart from making a mistake that led to at least eight subsequent recalculations of the number) was to name it γ (we have seen that Euler originally used C, and O and A have also been used). By such serendipity, its full accepted name is the Euler–Mascheroni constant. (A more distinguished legacy of Mascheroni is his result that any geometric construction that is possible with straight edge and compass can be achieved with a compass alone.)

Inevitably, things have moved on since then: in 1962 Donald Knuth took 250 terms of the Euler–Maclaurin series, with $n = 10\,000$ to compute γ to 1271 decimal places and in 1997 Thomas Papanikolaou computed it to 1 000 000 decimal places (the one millionth digit is 9) and in 1999 it was calculated to 108 000 000 decimal places by P. Demichel and X. Gourdon! Of course, such accuracy is far beyond anything that can conceivably prove 'useful', but that is not the point, an observation made in 1915 by James Glaisher (1848–1928) when he expressed the view:

> No doubt the desire to obtain the values of these quantities to a great many figures is also partly due to the fact that most of them are interesting in themselves; for e, π, γ, $\ln 2$, and many other numerical quantities occupy a curious and some of them almost a mysterious, place in mathematics, so that there is a natural tendency to do that can be done towards their precise determination.

Gamma as a Fraction

A man is like a fraction whose numerator is what he is and whose denominator is what he thinks of himself. The larger the denominator the smaller the fraction.

Count Lev Nikolgevich Tolstoy (1828–1910)

11.1 A MYSTERY

It is a simple matter of arithmetic to use the decimal approximations of a number to generate fractional approximations of it. For example,

$$\gamma = 0.577\,215\,664\,901\,532\,860\,606\,5 \ldots$$

results in the approximations:

$$\frac{5}{10}, \frac{57}{100}, \frac{577}{1000}, \frac{5772}{10\,000}, \frac{57\,721}{100\,000}, \cdots = \frac{1}{2}, \frac{57}{100}, \frac{577}{1000}, \frac{2881}{5000}, \frac{57\,721}{100\,000}, \cdots$$

Yet, compare the accuracy of the approximations with the mysterious sequence

$$\frac{3}{5}, \frac{4}{7}, \frac{11}{19}, \frac{15}{26}, \frac{71}{123}, \frac{228}{395}, \frac{3035}{5258}, \cdots$$

And what about $\frac{323\,007}{559\,595}$? These perplexing numbers are progressively more accurate approximations to γ and better than any comparable fraction arising as above. If we do want to approximate γ by fractions, we would do well to look to them. The question is, where do they come from?

11.2 A CHALLENGE

Fermat was given to posing number-theoretic problems. The most famous of them is his 'Last Theorem' (so called because it is the last of his assertions to succumb to proof), but there were numerous others. Euler disposed of many of them and one in particular was partly solved by him in 1759 and completed by Joseph-Louis Lagrange (1736–1813) in 1768. It was half of a challenge thrown to the European mathematical community by Fermat in January 1657 and read, 'Find a cube which, when increased by the sum of its proper divisors, becomes a square'; the other half was the same question with the words 'square' and 'cube' reversed. Bernard Frénicle de Bessy (1605–1675), an official at the French mint, a fine amateur mathematician and computor and correspondent of several of the great mathematical names of the time (particularly Fermat), provided four solutions to the first problem on the day he received it, and six more the day following. The challenge echoed across the English Channel to find the deaf ears of Wallis (who may well have been its main target) and the comment, 'Whatever the details of the matter, it finds me too absorbed by numerous occupations for me to be able to devote my attention to it immediately...'. Undeterred, a second challenge followed the next month, part of which was to find an integer y which would make $dy^2 + 1$ a perfect square for any positive integer d, or failing that, to solve the two special cases $d = 61, 109$. Again, Frénicle de Bessy played his part by calculating the smallest solutions for all $d < 150$ and challenged others to at least solve the cases $d = 150$ and $d = 313$, hinting that the second example may be beyond anyone's ability! Fermat fuelled the intellectual ferment with 'We await these solutions, which, if England or Belgic or Celtic Gaul do not produce, then Narbonese Gaul will.' (Narbonese Gaul was the area around Toulouse where Fermat lived.) Finally, rising to the bait, Wallis found particular solutions to both in very quick time and in doing so approached the solution of the ignored, initial challenge, as we show below. The challenges had generated interest in a problem that was 500 years older than Fermat and which became the subject of study and learned treatise by many, including the first president of The Royal Society, William Brouncker (1620?–1684).

If we consider the first challenge and make the reasonable assumption that Fermat had meant the cube to be that of a prime number, we require $1 + p + p^2 + p^3 = q^2$ or $(1 + p)(1 + p^2) = q^2$. Since 2 and only 2 (as the reader may wish to prove) is a factor of both brackets, the equation may be written as

$$ab = (\tfrac{1}{2}q)^2$$

with a and b co-prime.

Since a and b have no common factors we can legitimately conclude that $a = m^2$ and $b = n^2$ for some integers m and n and therefore that

$$1 + p = 2a = 2m^2 \quad \text{and} \quad 1 + p^2 = 2b = 2n^2,$$

so any such p must satisfy both the equations $p = 2m^2 - 1$ and $p^2 = 2n^2 - 1$. We are looking for primes of the form $2m^2 - 1$ whose squares are of the form $2n^2 - 1$, which looks as though it might be a big ask.

With this analysis we can see that the two challenges are essentially the same. The second equation is the more demanding of the two and is a special case of the problem: for a given non-square integer d find all x and y so that $x^2 = dy^2 \pm 1$, which is known as Pell's equation, after John Pell (1611–1685), another of the Founder Members of The Royal Society. His 1659 translation into English of one Johann Rahn's *Teutsche Algebra* brought to the English-speaking mathematical world the use of $\div$ for division and it may have been Pell himself who originated this use of the notation (the 'obelus' had been used for subtraction long before this). It was Euler who attributed Pell's name to the equation, but it is generally considered to be a rather generous (or mistaken) honour. With the plus sign and $d = 4\,729\,494$ it can also feature in the solution of a surprisingly difficult problem regarding the size of a herd of cattle, which was purportedly set by Archimedes to Apollonius as another (possibly revengeful) intellectual challenge. Whether or not Archimedes originally formulated the problem as a challenge or otherwise, it appeared in *The Sandreckoner*, which we mentioned on p. 3. It has subsequently earned the name of 'Archimedes' Revenge', as the herd turns out to have a size which has 206 545 digits.

11.3 AN ANSWER

What has this to do with those mysterious fractions that approximate γ (and any other number) so well? They are called the 'convergents' of what are known as 'continued fractions' (or, archaically, 'anthyphairetic ratios'). Firstly, it was Wallis who coined the name (in the 1653 edition of his book *Arithmatica Infinitorum*); they have been studied by any number of mathematicians over the years, including the 6th-century Indian mathematician Aryabhata (in whose work they make their first appearance), Johann Lambert and Joseph-Louis Lagrange (who made significant contributions to the theory), Christian Huygens (who used them in his design of a mechanical model of the Solar System), Euler (who laid down much of the modern theory of them, and used them to prove that both e and e^2 are irrational) and Gauss (who explored many of their deep properties). Perhaps their heyday was in the 19th century but there is a current resurgence of interest in them, partly through their connection with chaos theory and computer algorithms and they do have their part to play in our story. We will only see a tiny part of the use of this comparatively overlooked area of mathematics, but enough to be clear that they are more important than they at first seem and less difficult to use than they first look. Firstly, their definition.

A continued fraction is an expression of the form

$$a_0 + \cfrac{1}{a_1 + \cfrac{1}{a_2 + \cfrac{1}{a_3 + \cfrac{1}{a_4 + \cdots}}}},$$

where a_0 is an integer (possibly negative or zero) and a_1, a_2, ..., are non-zero positive integers; the expression could be finite or it could go on forever. Standard fractional notation is cumbersome and has given way to the alternative $[a_0; a_1, a_2, \ldots]$, in which the semi-colon separates the number's integer from its fractional part and the commas separate what are known as its 'partial quotients'. For example,

$$3 + \cfrac{1}{2 + \cfrac{1}{5 + \frac{1}{4}}} = 3 + \cfrac{1}{2 + \cfrac{1}{\left(\frac{21}{4}\right)}}$$

$$= 3 + \cfrac{1}{2 + \frac{4}{21}} = 3 + \cfrac{1}{\left(\frac{46}{21}\right)}$$

$$= 3 + \frac{21}{46} = \frac{159}{46}$$

or in a more compact notation, $[3; 2, 5, 4] = \frac{159}{46}$. If we build up the expression one term at a time, we get

$$3 + \frac{1}{2} = \frac{7}{2} \quad \text{and} \quad 3 + \cfrac{1}{2 + \frac{1}{5}} = \frac{38}{11},$$

thereby generating the 'convergents' of the partial fraction. Put another way, $\frac{159}{46}$ is approximately $\frac{7}{2}$ and also $\frac{38}{11}$, with the latter the better approximation. Clearly, any finite continued fraction can be telescoped into an ordinary fraction in this way, with each of the convergents successively better approximants to that fraction. Converting an ordinary fraction to its continued form simply requires us to strip off the integer part, invert and repeat the process; for example,

$$\frac{18}{13} = 1 + \frac{5}{13} = 1 + \cfrac{1}{\left(\frac{13}{5}\right)}$$

$$= 1 + \cfrac{1}{2 + \frac{3}{5}} = 1 + \cfrac{1}{2 + \left(\frac{5}{3}\right)}$$

$$= 1 + \cfrac{1}{2 + \cfrac{1}{1 + \frac{2}{3}}} = 1 + \cfrac{1}{2 + \cfrac{1}{1 + \left(\frac{3}{2}\right)}}$$

$$= 1 + \cfrac{1}{2 + \cfrac{1}{1 + \cfrac{1}{\left(1 + \frac{1}{2}\right)}}}$$

or $[1; 2, 1, 1, 2]$ and in the same way, $\frac{18}{13}$ is successively (and more accurately) approximated by $\frac{3}{2}$, $\frac{5}{3}$ and $\frac{7}{5}$. This highlights a possible source of ambiguity, as the $\frac{1}{2}$ above could have been inverted to 2 and then split into two 1s, but it is standard practice to agree that the fraction does not end with a 1.

11.4 THREE RESULTS

Continued fractions have many properties and are a fascinating subject in their own right, but at present we must resist the temptation to study them beyond mentioning the three properties of them that we will need, and even these we will not prove.

1. Each convergent is automatically in its lowest terms.

2. If p_n/q_n are convergents that approximate an irrational number x and if $q \leqslant q_n$ and if $p/q \neq p_n/q_n$, then $|p_n/q_n - x| < |p/q - x|$ and, more strongly, $|p_n - q_n x| < |p - qx|$.

   This means that each convergent of a continued fraction is the best-possible fractional approximation to x with a denominator of its size or less.

3. If x is an irrational number and a and b co-prime integers such that

$$\left| x - \frac{a}{b} \right| < \frac{1}{2b^2},$$

   then a/b is one of the convergents of the continued-fraction representation of x.

11.5 IRRATIONALS

The process of converting an irrational number to a continued fraction simply requires the decimal expansion to be dealt with in much the same way as a rational number. For example,

$$\pi = 3 + 0.141\,59\cdots = 3 + \cfrac{1}{7.062\,513\ldots} = 3 + \cfrac{1}{7 + \cfrac{1}{15.996\,594\ldots}}$$

$$= 3 + \cfrac{1}{7 + \cfrac{1}{15 + \cfrac{1}{1.003\,417\ldots}}}$$

$$= 3 + \cfrac{1}{7 + \cfrac{1}{15 + \cfrac{1}{1 + \cfrac{1}{292 + 0.654\ldots}}}},$$

which continues as

$$\pi = [3; 7, 15, 1, 292, 1, 1, 1, 2, 1, 3, 1, 14, 2, 1, 1, 2, 2, 2, 2, 1, 84, \dots],$$

with initial convergents $\frac{22}{7}, \frac{333}{106}, \frac{355}{113}$ and $\frac{103\,993}{33\,102}$.

Of course, $\frac{22}{7}$ is the approximation with which we are most familiar and, in what is possibly the first recorded attempt to approximate π, Archimedes included in his work *Measurement of a Circle* the bounds $\frac{223}{71} < \pi < \frac{22}{7}$, which he found by inscribing and circumscribing a circle with regular polygons of 96 sides. We know that $\frac{22}{7}$ is universally accepted as the most convenient approximation to use and with good reason, since we know from above that there is no fraction with a smaller denominator that is better. For the same reason, the more accurate $\frac{333}{106}$ is the best-possible rational approximant to π with denominator $\leqslant 106$, which says good things about the 16th-century European mathematicians who were known to use it and even better things about the Chinese mathematician Tsu Chung-chih (A.D. 430–501), who described $\frac{22}{7}$ as an 'inaccurate value' and $\frac{355}{113}$ as the 'accurate value' of π. Notice that the other Archimedean bound is not a convergent.

(It is impossible to resist mentioning the nice result that

$$\int_0^1 \frac{x^4(1-x)^4}{1+x^2}\,\mathrm{d}x = \frac{22}{7} - \pi,$$

which can be proved by using polynomial division and term-by-term integration to arrive at the indefinite integral $\frac{1}{7}x^7 - \frac{2}{3}x^6 + x^5 - \frac{4}{3}x^3 + 4x - 4\tan^{-1}x$.)

The continued fraction for other numbers can be found in the same way. For example, $\sqrt{2} = [1; 2, 2, 2, 2, \dots]$ with convergents $\frac{3}{2}, \frac{7}{5}, \frac{10}{7} \dots$.

The 'Golden Ratio'

$$\varphi = \tfrac{1}{2}(1 + \sqrt{5}) = [1; 1, 1, 1, 1, \dots],$$

with convergents the Fibonacci numbers

$$\tfrac{2}{1}, \tfrac{3}{2}, \tfrac{5}{3}, \dots,$$

$$e = [2; 1, 2, 1, 1, 4, 1, 1, 6, 1, 1, 8, 1, 1, 10, 1, 1, 12, \dots],$$

with convergents $\frac{5}{2}, \frac{8}{3}, \frac{11}{4}, \frac{19}{7}, \frac{73}{32}, \dots$.

Notice how the continued-fraction representation of these numbers reveals an otherwise hidden pattern and one that makes them exceptional in an important and strange way, which we will discuss in Chapter 14.

It is also true that $\pi^4 = [97; 2, 2, 2, 2, 16\,539, 1, \dots]$, which makes the fifth convergent $\frac{35\,444\,733}{363\,875}$ a particularly accurate rational approximation to π^4 (and therefore its fourth root is a particularly accurate decimal approximation to π—differing in the 13th decimal place).

Those earlier fractional approximations to γ come, of course, from its own continued fraction form of

$$\gamma = [0; 1, 1, 2, 1, 2, 1, 4, 3, 13, 5, 1, 1, 8, 1, 2, 4, 1, 1,$$
$$40, 1, 11, 3, 7, 1, 7, 1, 1, 5, 1, 49, 4, 1, 65, \dots]$$

with convergents

$$1, \frac{1}{2}, \frac{3}{5}, \frac{4}{7}, \frac{11}{19}, \frac{15}{26}, \frac{71}{123}, \frac{228}{395}, \frac{3035}{5258}, \dots, \frac{323\,007}{559\,595}, \dots$$

As an indication of the accuracy that is achieved,

$$\left| \frac{323\,007}{559\,595} - \gamma \right| = 1.025 \times 10^{-12}.$$

Thomas Papanikolaou, who was mentioned earlier, also calculated the continued fraction for γ up to and including the 470 006th partial quotient and from this he could conclude that if γ is rational, the denominator of the fraction must be greater than $10^{242\,080}$. Of course, an infinite number of fractions with such denominators (and larger) do exist, but (unreliable) intuition moves us to think that a 'naturally occurring' number, such as γ, simply would not behave in such an extreme way; to confound that view, someone needs to produce an accepted 'natural' fraction with such a denominator! This aspect of γ's behaviour was touched on by the great German mathematician David Hilbert (1862–1943) in a seminal lecture given in 1900, which we will describe in more detail and from which we will quote at greater length later:

> Take any definite unsolved problem, such as the question as to the irrationality of the Euler–Mascheroni constant C, or the existence of an infinite number of prime numbers of the form $2^n + 1$. However, unapproachable these problems may seem to us and however helpless we stand before them, we have, nevertheless, the firm conviction that their solution must follow by a finite number of purely logical processes.

The mathematical world still awaits the discovery of that particular 'finite number of purely logical processes'.

The connection with Pell's equation is profound.

11.6 PELL'S EQUATION SOLVED

The solutions to Pell's equation are hardly predictable: if we take it as $a^2 - db^2 = 1$, then with $d = 60$ the smallest solution is $a = 31$, $b = 4$; with $d = 62$ it is $a = 63$, $b = 8$ yet with $d = 61$ it is $a = 1\,766\,319\,049$, $b = 226\,153\,980$!

If a and b satisfy $a^2 - db^2 = 1$, they cannot possibly have any common factors. That said, the underlying pattern is revealed by the following argument:

$$a^2 - db^2 = 1 \Leftrightarrow (a - b\sqrt{d})(a + b\sqrt{d}) = 1 \Leftrightarrow \frac{a}{b} - \sqrt{d} = \frac{1}{b(a + b\sqrt{d})}.$$

The factorization makes clear that $a > b\sqrt{d}$ and so we can manufacture the inequality

$$0 < \frac{a}{b} - \sqrt{d} < \frac{\sqrt{d}}{b(b\sqrt{d} + b\sqrt{d})} = \frac{\sqrt{d}}{2b^2\sqrt{d}} = \frac{1}{2b^2}.$$

Invoke the third property of continued fractions and we see that a/b must be a convergent of $\sqrt{d}$, so the search for the solutions of Fermat's problems should be among the continued-fraction expansion of the number defining them. For example, with the first problem, the first solution is $p = 7$ and $n = 5$ to give $p = 7$ and $q = 20$ as the smallest solution.

11.7 FILLING THE GAPS

Continued fractions are the first choice among many possibilities for rational approximation, but they do leave plenty of gaps in the list of best possible approximants. With γ we have seen that we have the consecutive continued fraction convergents of

$$1, \frac{1}{2}, \frac{3}{5}, \frac{4}{7}, \frac{11}{19}, \frac{15}{26}, \frac{71}{123}, \frac{228}{395}, \frac{3035}{5258}, \ldots,$$

but if we set a computer to search for the best-possible rational approximations up to and including any given denominator, we get

$$1, \frac{1}{2}, \frac{3}{5}, \frac{4}{7}, \frac{11}{19}, \frac{15}{26}, \frac{41}{71}, \frac{56}{97}, \frac{71}{123}, \frac{157}{272}, \frac{228}{395}, \ldots$$

and the next interval is filled with

$$\ldots, \frac{228}{395}, \frac{1667}{2888}, \frac{1895}{3283}, \frac{2123}{3678}, \frac{2351}{4073}, \frac{2579}{4468}, \frac{2807}{4863}, \frac{3035}{5258}, \ldots$$

and, of course, the gaps get bigger and so does the list of fractions to fill them.

In short, continued fractions provide a nice, methodical method of rational approximation and they are extremely useful in general theory, but they do not tell the whole story; very few things do.

11.8 THE HARMONIC ALTERNATIVE

We will introduce (without pursuing the idea far) just one alternative method of fractional approximation, mainly because it encourages deeper thought into our base 10 system and also it provides a first example of the usefulness of the terms of the harmonic series. It has to be said that the method does not find the best fractions, but it is nonetheless novel and worthy of study.

We are really interested in the fractional part of a number and so we will choose to divide a decimal fraction into its whole number part, considered as a single number, and its fractional part, divided into its components. For example, the expression 62.372 58 is a shorthand for

$$62 + \frac{1}{10} \times 3 + \frac{1}{10^2} \times 7 + \frac{1}{10^3} \times 2 + \frac{1}{10^4} \times 5 + \frac{1}{10^5} \times 8,$$

which can be written in the rather more complicated form

$$62 + \frac{1}{10}\left(3 + \frac{1}{10}\left(7 + \frac{1}{10}\left(2 + \frac{1}{10}\left(5 + \frac{1}{10}(8)\right)\right)\right)\right)$$

and of course such expressions could be extended indefinitely, as the number dictates. The 3, 7, 2, 5, 8 are simply a special case of any sequence of non-negative integers which each less than 10 and we could adopt notation similar to that for continued fractions, writing the number as $[62; 3, 7, 2, 5, 8]$. More generally,

$$[n; a, b, c, \ldots] = n + \frac{1}{10}\left(a + \frac{1}{10}\left(b + \frac{1}{10}\left(c + \cdots\right)\right)\right),$$

where n is the whole number part and $a, b, c, \ldots$ form the fractional part and so are non-negative integers $\leqslant 9$.

So far this is doing no more than looking at the obvious in a different way and playing with notation, but the expanded form of the expression, with its repeated $\frac{1}{10}$, suggests that we could alter that number to a different one to achieve a representation in another base (the $a, b, c, \ldots$ would naturally be restricted to be less than that base). This is nothing new. Replace $\frac{1}{10}$ by $\frac{1}{2}$ and we have the binary system of 0s and 1s, with $\frac{1}{3}$ the tertiary system, etc. More interesting still, what if we mix the bases and represent the number in a mixed-base system—using the terms of the harmonic series? This would mean that our number would be written in the form

$$n + \frac{1}{2}\left(a + \frac{1}{3}\left(b + \frac{1}{4}\left(c + \cdots\right)\right)\right)$$

and rational approximations to it would be any first part of this, where $a < 2$, $b < 3, c < 4, \ldots.$

A closer look at the form of this representation reveals that, rather than writing the number as

$$n + \frac{1}{10}a + \frac{1}{10^2}b + \cdots \quad \text{with } a, b, \cdots < 9,$$

we are writing it as

$$n + \frac{1}{2!}a + \frac{1}{3!}b + \frac{1}{4!}c + \cdots \quad \text{with } a < 2, \ b < 3, \ c < 4, \ldots.$$

If we start with π, we get

$$\pi = 3 + \frac{1}{2}\left(0 + \frac{1}{3}\left(0 + \frac{1}{4}\left(3 + \frac{1}{5}\left(1 + \frac{1}{6}\left(5 + \frac{1}{7}\left(6 + \frac{1}{8}\left(5 + \cdots\right)\right)\right)\right)\right)\right)\right)$$

or in the more compact notation

$$\pi = [3; 0, 0, 3, 1, 5, 6, 5, \ldots]$$

to give the fractional approximations

$$3, 3, 3, \frac{25}{8}, \frac{47}{15}, \frac{2261}{720}, \frac{15\,833}{5040}, \frac{42\,223}{13\,440}, \frac{11\,400\,211}{3\,628\,800}, \ldots$$

Since the Taylor expansion of e^x is

$$1 + x + \frac{x^2}{2!} + \frac{x^3}{3!} + \frac{x^4}{4!} + \cdots,$$

putting $x = 1$ gives e as

$$1 + 1 + \frac{1}{2}\left(1 + \frac{1}{3}\left(1 + \frac{1}{4}\left(1 + \frac{1}{5}\left(1 + \frac{1}{6}\left(1 + \frac{1}{7}\left(1 + \frac{1}{8}\left(1 + \cdots\right)\right)\right)\right)\right)\right)\right)$$

or the very nice $e = [2; 1, 1, 1, 1, 1, 1, 1, \ldots]$ to give fractional approximations

$$\frac{5}{2}, \frac{8}{3}, \frac{65}{24}, \frac{163}{60}, \frac{1957}{720}, \frac{6855}{252}, \frac{109\,601}{40320}, \ldots$$

Finally, with γ we have

$$\gamma = 0 + \frac{1}{2}\left(1 + \frac{1}{3}\left(0 + \frac{1}{4}\left(1 + \frac{1}{5}\left(4 + \frac{1}{6}\left(1\right.\right.\right.\right.\right.$$
$$\left.\left.\left.\left.\left. + \frac{1}{7}\left(4 + \frac{1}{8}\left(1 + \frac{1}{9}\left(3 + \frac{1}{10}0\right) \cdots\right)\right)\right)\right)\right)\right)$$

or in the shorthand notation $[0; 0, 1, 0, 1, 4, 1, 4, 1, 3, 0, \ldots]$ and the rational approximations

$$\frac{1}{2}, \frac{1}{2}, \frac{13}{24}, \frac{23}{40}, \frac{83}{144}, \frac{2909}{5040}, \frac{23\,273}{40\,320}, \frac{3491}{6048}, \frac{3491}{6048}, \ldots$$

These various approximations are not at all bad, as the reader can measure. Notice that with the possibility of a zero, consecutive approximations can be the same. We will soon be looking at a variety of other ways in which the harmonic series appears.

Where Is Gamma?

One cannot escape the feeling that these mathematical formulas have an independent existence and an intelligence of their own, that they are wiser than we are, wiser even than their discoverers, that we get more out of them than was originally put into them.

Heinrich Hertz (1857–1894)

Gamma's definition $\gamma = \lim_{n\to\infty}(H_n - \ln n)$, when rewritten as the asymptotic approximation $H_n \approx \ln n + \gamma$, provides a simple (and accurate) method for approximating the partial sums of the harmonic series. The lack of an explicit formula for H_n together with its glacially slow divergence makes the approximation all the more important and with that approximation we have an inevitable appearance of γ; already we have seen the estimate used on a number of occasions. Its connection with the Gamma function guarantees γ's role in analysis and the Gamma function's connection with the Zeta functions guarantees γ's role in number theory. The number is inevitably, intrinsically (and frequently, intricately) involved in mathematics, reticent though it is to show itself in elementary areas of the subject. It would be easy to relegate this chapter to a long list of integrals, sums, products and limits which involve γ but instead we will give a representative few and leave it to the interested reader to seek out more; in doing this we will be paying no more than lip-service to that 'serious consideration' of which it is worthy. To begin with we will look at another example of γ allowing the harmonic series to be replaced by logarithms, this time not as an estimate but as the exact limit.

12.1 THE ALTERNATING HARMONIC SERIES REVISITED

The name Riemann has already appeared several times, attached to the word Hypothesis. It is not yet time to consider either the man or the problem but we can now mention a peculiar result of his regarding the convergence of series and its novel implications for γ (and any other number).

The (geometric) series $1 - \frac{1}{2} + \frac{1}{4} - \frac{1}{8} \cdots$ converges to $\frac{2}{3}$ and the series of positive terms associated with it $1 + \frac{1}{2} + \frac{1}{4} + \frac{1}{8} \cdots$ to 2. Yet, it is not always the case that sacrificing the cancellation brought about by omitting the minus signs has such innocent consequences, and the harmonic series is a particular case in point: we know that the alternating harmonic series converges (to ln 2) and that the harmonic series diverges. This phenomenon is encapsulated in the concept of 'conditional convergence', with the alternating harmonic series conditionally convergent and that alternating geometric series above 'absolutely convergent'. Conditionally convergent series are delicate, as we can see from

$$1 - \frac{1}{2} - \frac{1}{4} + \frac{1}{3} - \frac{1}{6} - \frac{1}{8} + \frac{1}{5} - \frac{1}{10} - \frac{1}{12} + \frac{1}{7} - \frac{1}{14} - \frac{1}{16} + \cdots$$

$$= \left(1 - \frac{1}{2}\right) - \frac{1}{4} + \left(\frac{1}{3} - \frac{1}{6}\right) - \frac{1}{8} + \left(\frac{1}{5} - \frac{1}{10}\right)$$

$$- \frac{1}{12} + \left(\frac{1}{7} - \frac{1}{14}\right) - \frac{1}{16} + \cdots$$

$$= \frac{1}{2} - \frac{1}{4} + \frac{1}{6} - \frac{1}{8} + \frac{1}{10} - \frac{1}{12} + \frac{1}{14} - \frac{1}{16} + \cdots$$

$$= \frac{1}{2}\left(1 - \frac{1}{2} + \frac{1}{3} - \frac{1}{4} + \frac{1}{5} - \frac{1}{6} + \frac{1}{7} - \frac{1}{8} + \cdots\right)$$

$$= \frac{1}{2}\ln 2$$

and the alternating harmonic series now converges to half of itself!

Riemann's peculiar result is that any conditionally convergent series can be made to sum to any number at all! For example, if we wish an arrangement of the alternating harmonic series to sum to the Golden Ratio $\varphi = \frac{1}{2}(1 + \sqrt{5})$, that arrangement begins

$$\varphi = 1 + \frac{1}{3} + \frac{1}{5} + \frac{1}{7} - \frac{1}{2} + \frac{1}{9} + \frac{1}{11} + \frac{1}{13} + \frac{1}{15} + \frac{1}{17} + \frac{1}{19} - \frac{1}{4} + \cdots$$

$$+ \frac{1}{21} + \frac{1}{23} + \frac{1}{25} + \frac{1}{27} + \frac{1}{29} + \frac{1}{31} - \frac{1}{6} + \cdots.$$

We can manufacture an arrangement to sum to any given number l by adding in as many of the positive, odd terms as are needed to make the sum exceed l, bring in the negative even terms to bring the sum below l, and continue in this see-saw way for as long as we please; the divergence of each of the two subseries guarantees that we will always be able to do this.

There are general results associated with this phenomenon, the proof of one of which naturally brings in (and takes out) γ and to look at it we need the concept of a 'simple' arrangement of the alternating harmonic series. Such an arrangement is defined to be any in which the terms of the two subsequences of positive and of negative terms appear in descending order. For example, the re-arrangements which led to $\frac{1}{2}\ln 2$ and to φ are both simple, yet the rearrangement $1 + \frac{1}{2} - \frac{1}{3} + \frac{1}{6} - \frac{1}{5} + \frac{1}{4} - \cdots$ is not.

With this definition in place, let p_n be the number of positive and q_n the number of negative terms in the first n terms of a simply rearranged alternating harmonic series, then the result to interest us is that the rearrangement converges if and only if

$$\alpha = \lim_{n \to \infty} \frac{p_n}{q_n}$$

exists, in which case the sum is $\ln 2 + \frac{1}{2} \ln \alpha$. Above we have $\alpha = \frac{1}{2}$ and so the sum is $\ln 2 + \frac{1}{2} \ln \frac{1}{2} = \frac{1}{2} \ln 2$.

To establish the result, write the sum of the first n terms of the series as $\sum_{k=1}^{n} a_k$, then

$$\sum_{k=1}^{n} a_k = \sum_{k=1}^{p_n} \frac{1}{2k-1} - \sum_{k=1}^{q_n} \frac{1}{2k}.$$

But

$$\sum_{k=1}^{p_n} \frac{1}{2k-1} = \sum_{k=1}^{2p_n} \frac{1}{k} - \sum_{k=1}^{p_n} \frac{1}{2k}$$

and so

$$\sum_{k=1}^{n} a_k = \sum_{k=1}^{2p_n} \frac{1}{k} - \sum_{k=1}^{p_n} \frac{1}{2k} - \sum_{k=1}^{q_n} \frac{1}{2k}$$

$$= H_{2p_n} - \tfrac{1}{2} H_{p_n} - \tfrac{1}{2} H_{q_n}$$

$$= (\ln 2p_n - \gamma_{2p_n}) - \tfrac{1}{2}(\ln p_n - \gamma_{p_n}) - \tfrac{1}{2}(\ln q_n - \gamma_{q_n})$$

$$= \ln 2 + \tfrac{1}{2} \ln \left(\frac{p_n}{q_n} \right) - \gamma_{2p_n} + \tfrac{1}{2}\gamma_{p_n} + \tfrac{1}{2}\gamma_{q_n},$$

where the γ_n are the approximations to γ to that number of terms.

Therefore,

$$\lim_{n \to \infty} \sum_{k=1}^{n} a_k = \ln 2 + \tfrac{1}{2} \ln \left(\lim_{n \to \infty} \frac{p_n}{q_n} \right) - \gamma + \tfrac{1}{2}\gamma + \tfrac{1}{2}\gamma$$

$$= \ln 2 + \tfrac{1}{2} \ln \alpha$$

and we are done.

If, for example, we wish to write $\ln 3$ in terms of the alternating harmonic series it must be that

$$\ln 3 = \ln 2 + \tfrac{1}{2} \ln \alpha,$$

which makes $\alpha = \frac{9}{4}$, and indeed the representation is

$$\ln 3 = 1\left(+\frac{1}{3} - \frac{1}{2} + \frac{1}{5} + \frac{1}{7} - \frac{1}{4} + \frac{1}{9} + \frac{1}{11}\right.$$

$$\left. -\frac{1}{6} + \frac{1}{13} + \frac{1}{15} - \frac{1}{8} + \frac{1}{17} + \frac{1}{19}\right)$$

$$\left(+\frac{1}{21} - \frac{1}{10} + \frac{1}{23} + \frac{1}{25} - \frac{1}{12} + \frac{1}{27}\right.$$

$$\left. +\frac{1}{29} - \frac{1}{14} + \frac{1}{31} + \frac{1}{33} - \frac{1}{16} + \frac{1}{35} + \frac{1}{37}\right)$$

$$\left(+\frac{1}{39} - \frac{1}{18} + \cdots \right) \cdots ,$$

where the bracketing groups equal numbers of the terms by matching patterns of the signs. Each group comprises nine $+$ signs and four $-$ signs and with this pattern repeated throughout the expansion we will have $\alpha = \frac{9}{4}$, as required.

Of course α is the limit of p_n/q_n and we cannot expect in general the limit to reveal itself by simple repetition. Recall that Euler had hoped that γ might be the logarithm of some important number. If that is the case, it would be possible to write

$$\gamma = \ln 2 + \tfrac{1}{2}\ln\alpha,$$

where the α is the 'important' limit of the ratio of the $+$ and $-$ signs it its representation in terms of the alternating harmonic series, which starts

$$\gamma = 1 - \frac{1}{2}$$

$$+\frac{1}{3}\left\{\left(-\frac{1}{4} - \frac{1}{6} + \frac{1}{5} - \frac{1}{8} + \frac{1}{7} - \frac{1}{10} + \frac{1}{9} - \frac{1}{12} + \frac{1}{11}\right)\right.$$

$$\left(-\frac{1}{14} - \frac{1}{16} + \frac{1}{13} - \frac{1}{18} + \frac{1}{15} - \frac{1}{20} + \frac{1}{17} - \frac{1}{22} + \frac{1}{19}\right)$$

$$\left(-\frac{1}{24} - \frac{1}{26} + \frac{1}{21} - \frac{1}{28} + \frac{1}{23} - \frac{1}{30} + \frac{1}{25} - \frac{1}{32} + \frac{1}{27}\right)$$

$$\left(-\frac{1}{34} - \frac{1}{36} + \frac{1}{29} - \frac{1}{38} + \frac{1}{31} - \frac{1}{40} + \frac{1}{33} - \frac{1}{42} + \frac{1}{35}\right)$$

$$\left(-\frac{1}{44} - \frac{1}{46} + \frac{1}{37} - \frac{1}{48} + \frac{1}{39} - \frac{1}{50} + \frac{1}{41} - \frac{1}{52} + \frac{1}{43}\right)$$

$$\left.\left(-\frac{1}{54} - \frac{1}{56} + \frac{1}{45} - \frac{1}{58} + \frac{1}{47} - \frac{1}{60} + \frac{1}{49}\right)\right\}$$

$$\left\{\left(-\frac{1}{62} - \frac{1}{64} - \cdots\right)\cdots\right\}.$$

Within the braces there are five round brackets, each containing nine terms with identical sign patterns and, at the end, one bracket of seven terms. If this repeat is continued, we would have $23 +$ signs and $29 -$ signs in the repeated cycle of 52 terms; this would mean that $\alpha = \frac{23}{29}$ and $\gamma = \ln 2 + \frac{1}{2} \ln \frac{23}{29}$, and we will have outdone the great Euler! His constant is then

$$\gamma = \ln 2 \sqrt{\frac{23}{29}}.$$

Unfortunately, it isn't, since this evaluates to $0.577\,246.\ldots$. That cycle was too much to hope for and the pattern breaks at around the 550th term. Is there a pattern with a longer repetition? Who knows? If $\gamma = \ln 2 + \frac{1}{2} \ln \alpha$, then $\alpha = \frac{1}{4} e^{2\gamma}$ and the convergents of the continued fraction of this number are

$$\frac{3}{4}, \frac{4}{5}, \frac{19}{24}, \frac{23}{29}, \frac{548}{691}, \frac{571}{720}, \frac{1119}{1411}, \frac{2809}{3542}, \frac{6737}{8495}, \frac{63\,442}{79\,997}, \frac{450\,831}{568\,474}, \cdots$$

and at least we have hit on one of them, with our $\frac{23}{29}$ making an appearance.

12.2 IN ANALYSIS

One of the (many) problems with integration is that we cannot always integrate a function in 'closed form'; that is, no combination of the usual functions of mathematics will combine to be the anti-derivative of the function, and there is often only a slight change needed to convert possible to impossible, or the other way around. For example, $\ln u, u \ln u, (\ln u)/u, 1/(u \ln u)$ are all straightforward to integrate, yet $1/\ln u, u/\ln u$ are simply not possible. The irksome thing is that some of these 'difficult' integrals occur with great frequency and in many important applications, so much so that they lose their anonymity and are given names. For example,

$$\int \frac{\sin u}{u}\,du, \quad \int \frac{\cos u}{u}\,du, \quad \int e^{-u^2}\,du, \quad \int \frac{e^{-u}}{u}\,du, \quad \int \frac{1}{\ln u}\,du$$

are all impossible in closed form and give rise to the functions:

$$\mathrm{erf}(x) = \frac{2}{\sqrt{\pi}} \int_0^x e^{-u^2}\,du \qquad \text{(the error function)},$$

$$Li(x) = \int_2^x \frac{1}{\ln u}\,du \qquad \text{(the logarithmic integral)},$$

$$Ci(x) = \int_x^\infty \frac{\cos u}{u}\,du \qquad \text{(the cosine integral)},$$

105

$$Si(x) = \int_0^x \frac{\sin u}{u}\, du \qquad \text{(the sine integral),}$$

$$Ei(x) = \int_x^\infty \frac{e^{-u}}{u}\, du \qquad \text{(the exponential integral).}$$

They each appear in their different ways and places.

Laplace's error function, $\mathrm{erf}(x)$, is easily recognized as essentially the probability density function of the Normal distribution (the constant is needed to make the total area 1).

$Li(x)$ appears regularly in number theory in estimates of asymptotic values, including a conjecture of Littlewood and Hardy concerning the Goldbach Conjecture (mentioned in a few lines). It will later become the central focus of our attention when it appears as Gauss's estimate of the prime counting function $\pi(x)$—Euler had, of course, already considered the function (in 1768). Subsequently it appears in the work of Mascheroni (1790) and Caluso (1805) but it came to prominence (and was given its name) after it was the object of study in Soldner's *Theory of a New Transcendental Function* of 1809 (admittedly with the alternative lower limit of 0). We mentioned this work on p. 89. It was in that paper that Soldner gave that corrected value of γ and also the series expansion of $Li(x)$ as

$$Li(x) = \gamma + \ln \ln x + \sum_{r=1}^\infty \frac{\ln^r x}{rr!}.$$

$Ci(x)$ has the similar form

$$Ci(x) = -\gamma - \ln x - \sum_{r=1}^\infty \frac{(-x^2)^r}{2r(2r)!}$$

but $Si(x)$ involves neither $\ln$ nor γ in its expansion of

$$Si(x) = \sum_{r=1}^\infty (-1)^{r-1} \frac{x^{2r-1}}{(2r-1)(2r-1)!}.$$

This last trio work hand-in-hand in many applications and in widely diverse areas of mathematics, including quantum field theory, electromagnetic theory, semiconductor physics, and analysis of the Gibbs phenomena of Fourier analysis (the misbehaved bits at the fly-back points).

$Ei(x)$ is important partly because the integral of any function of the form $R(x)e^x$, where $R(x)$ is a rational function, can be shown to reduce to elementary integrals and $Ei(x)$.

γ also appears in what are known as 'modified Bessel functions of the second kind', named after the German astronomer F. W. Bessel (1784–1846), although they were studied earlier by yet another Bernoulli (Daniel) (1700–1782) and, inevitably, Euler. These functions appear among the solutions of what is known as the Bessel Equation,

$$x^2\frac{d^2y}{dx^2} + x\frac{dy}{dx} + (x^2 - \alpha^2)y = 0,$$

where $\alpha \geq 0$ is constant. It arises in the study problems concerning vibrations of membranes, heat flow in cylinders and the propagation of electric currents in cylindrical conductors—and some of the problems of analytic number theory.

Other, nameless integrals and limits involving γ are easy to find. Recall that $\gamma = -\Gamma'(1)$ and we have, using one of the definitions of the Gamma function (and differentiating under the integral sign),

$$\Gamma(x) = \int_0^\infty u^{x-1}e^{-u}\,du = \int_0^\infty e^{(x-1)\ln u}e^{-u}\,du$$

so

$$\Gamma'(x) = \int_0^\infty u^{x-1}e^{-u}\ln u\,du \quad \text{and} \quad \Gamma'(1) = \int_0^\infty e^{-u}\ln u\,du,$$

which makes

$$\gamma = -\int_0^\infty e^{-u}\ln u\,du.$$

Increasing the level of ingenuity develops this into a more exotic result:

$$-\gamma = \int_0^\infty e^{-u}\ln u\,du = \int_0^1 e^{-u}\ln u\,du + \int_1^\infty e^{-u}\ln u\,du.$$

Now we integrate by parts in each case, the second integral perfectly straightforwardly, integrating e^{-u} to $-e^{-u}$, but the first by using the underhand trick of integrating e^{-u} to $-e^{-u} + 1$ to get

$$-\gamma = [(-e^{-u} + 1)\ln u]_0^1$$
$$-\int_0^1 \frac{(-e^{-u} + 1)}{u}\,du + [-e^{-u}\ln u]_1^\infty - \int_1^\infty \frac{-e^{-u}}{u}\,du.$$

The two evaluated components are both 0, with the exponential drowning the logarithm, and so

$$-\gamma = -\int_0^1 \frac{(-e^{-u} + 1)}{u}\,du + \int_1^\infty \frac{e^{-u}}{u}\,du$$

and

$$\gamma = \int_0^1 \frac{(1 - e^{-u})}{u}\,du - \int_1^\infty \frac{e^{-u}}{u}\,du.$$

107

Finally, if we make the substitution $u = 1/t$ in the second integral and swap back variables to u we get

$$\gamma = \int_0^1 \frac{(1 - e^{-u})}{u}\, du - \int_0^1 \frac{e^{-1/u}}{u}\, du = \int_0^1 \frac{1 - e^{-u} - e^{-1/u}}{u}\, du.$$

This is not only a fearsome integral conquered but also an integral definition of γ over a finite interval, which can be used to calculate its value, provided that for reasons of continuity we agree to define the integrand to be 1 at $u = 0$ (graphing the function is a good test for any graph plotter and numerically approximating the area an even better one).

With this result we can derive the series expansion for the $Ei(x)$ function, using a method very similar to Soldner's for $Li(x)$.

Assume $x \geqslant 1$ (the ideas still work if $x < 1$), then

$$Ei(x) = \int_x^\infty \frac{e^{-u}}{u}\, du = \int_1^\infty \frac{e^{-u}}{u}\, du - \int_1^x \frac{e^{-u}}{u}\, du$$

$$= \int_1^\infty \frac{e^{-u}}{u}\, du - \int_1^x \frac{e^{-u} - 1}{u} + \frac{1}{u}\, du$$

$$= \int_1^\infty \frac{e^{-u}}{u}\, du - \int_1^x \frac{e^{-u} - 1}{u}\, du - \int_1^x \frac{1}{u}\, du$$

$$= \int_1^\infty \frac{e^{-u}}{u}\, du - \left(\int_0^x \frac{e^{-u} - 1}{u}\, du - \int_0^1 \frac{e^{-u} - 1}{u}\, du \right) - \int_1^x \frac{1}{u}\, du$$

$$= \int_1^\infty \frac{e^{-u}}{u}\, du - \int_0^x \frac{e^{-u} - 1}{u}\, du + \int_0^1 \frac{e^{-u} - 1}{u}\, du - \int_1^x \frac{1}{u}\, du$$

$$= \int_1^\infty \frac{e^{-u}}{u}\, du + \int_0^1 \frac{e^{-u} - 1}{u}\, du - \int_0^x \frac{e^{-u} - 1}{u}\, du - \int_1^x \frac{1}{u}\, du$$

$$= -\gamma - \int_0^x \sum_{r=1}^\infty (-1)^r \frac{u^{r-1}}{r!}\, du - \ln x$$

$$= -\gamma - \sum_{r=1}^\infty (-1)^r \frac{1}{r!} \int_0^x u^{r-1}\, du - \ln x$$

$$= -\gamma - \sum_{r=1}^\infty (-1)^r \frac{1}{r!} \left[\frac{u^r}{r} \right]_0^x - \ln x$$

$$= -\gamma - \sum_{r=1}^\infty (-1)^r \frac{x^r}{rr!} - \ln x = -\gamma - \ln x - \sum_{r=1}^\infty \frac{(-x)^r}{rr!}.$$

We have used the standard Taylor expansion of e^{-u} to deal with the third integral.

There are a countless number of other integrals, sums and products in which γ is involved and below we list a few more anonymous examples:

$$\int_0^1 \ln \ln \frac{1}{x}\, dx = -\gamma, \qquad \int_0^\infty e^{-x^2} \ln x\, dx = -\frac{\sqrt{\pi}}{4}(\gamma + 2\ln 2),$$

$$\int_0^1 \frac{1}{\ln x} + \frac{1}{1-x}\, dx = \gamma, \qquad \int_0^\infty e^{-x} \ln^2 x\, dx = \frac{\pi^2}{6} + \gamma^2,$$

$$\lim_{n\to\infty}\left(n - \Gamma\left(\frac{1}{n}\right)\right) = \gamma, \qquad \lim_{n\to\infty} \frac{1}{\ln n}\prod_{p\leqslant n}\left(1 - \frac{1}{p}\right)^{-1} = e^\gamma,$$

$$\lim_{x\to 1+}\sum_{n=1}^\infty \frac{1}{n^x} - \frac{1}{x^n} = \gamma, \qquad \lim_{n\to\infty}\frac{1}{\ln n}\prod_{p\leqslant n}\left(1 + \frac{1}{p}\right) = \frac{6e^\gamma}{\pi^2},$$

$$\sum_{r=2}^\infty \frac{\Lambda(r) - 1}{r} = -2\gamma, \qquad \int_0^\infty e^{-x}\left(\frac{1}{1 - e^{-x}} - \frac{1}{x}\right) dx = \gamma,$$

$$\sum_{i=2}^\infty \frac{1}{i}(\zeta(i) - 1) = 1 - \gamma, \qquad \int_1^\infty \frac{\{x\}}{x^2}\, dx = \int_1^\infty \frac{x - \lfloor x \rfloor}{x^2}\, dx = 1 - \gamma,$$

The two integrals evaluating to expressions involving π display nice relationships between it, e and γ. The two product forms were both arrived at in 1874 by Franz Mertens (1840–1927) and we will have use of one of them in Chapter 15. The p that appears is prime and the first form can be developed to the very nice

$$\gamma = \lim_{n\to\infty}\left\{ -\ln\ln n - \sum_{p\leqslant n}\ln\left(1 - \frac{1}{p}\right)\right\}$$

$$= \lim_{n\to\infty}\left\{ \sum_{p\leqslant n}\left(\frac{1}{p} + O\left(\frac{1}{p^2}\right)\right) - \ln\ln n\right\},$$

which is reminiscent of Gamma's definition, but using primes only.

The second summation result involves the Von Mangoldt function

$$\Lambda(r) = \begin{cases} \ln p, & r = p^m,\ p \text{ prime}, \\ 0, & \text{otherwise}, \end{cases}$$

which will come to our closer attention again in Chapter 16.

Each expression in the list is established in its own way and we will content ourselves with proving just two of them: the one involving the Floor function and the other the Zeta function, and so keep an earlier promise, made on p. 52.

Firstly we will deal with

$$\int_1^\infty \frac{\{x\}}{x^2}\, dx,$$

where the notation $\{x\}$ is used for the fractional part of x and is therefore related to the Floor function by $\{x\} = x - \lfloor x \rfloor$. It seems impossible ever to arrive at an exact answer for such a strange integral, but we will see that γ naturally makes an appearance which solves the problem.

To begin with, by definition of the Floor function,

$$\int_1^n \frac{\lfloor x \rfloor}{x^2} \, dx = \int_1^n \frac{1}{x^2} \left(\sum_{1 \leqslant r \leqslant x} 1 \right) dx.$$

We need to rewrite this expression and to do so imagine the interval divided into unit sub-intervals with the right-hand-side end point excluded, then

$$\int_1^n \frac{1}{x^2} \left(\sum_{1 \leqslant r \leqslant x} 1 \right) dx$$

$$= \int_1^2 \frac{1}{x^2} \left(\sum_{1 \leqslant r \leqslant x} 1 \right) dx + \int_2^3 \frac{1}{x^2} \left(\sum_{1 \leqslant r \leqslant x} 1 \right) dx$$

$$+ \int_3^4 \frac{1}{x^2} \left(\sum_{1 \leqslant r \leqslant x} 1 \right) dx + \cdots + \int_{n-1}^n \frac{1}{x^2} \left(\sum_{1 \leqslant r \leqslant x} 1 \right) dx$$

$$= \int_1^2 \frac{1}{x^2} (1) \, dx + \int_2^3 \frac{1}{x^2} (2) \, dx$$

$$+ \int_3^4 \frac{1}{x^2} (3) \, dx + \cdots + \int_{n-1}^n \frac{1}{x^2} (n-1) \, dx. \qquad (12.1)$$

Now consider the expression

$$\sum_{1 \leqslant r \leqslant n} \int_r^n \frac{1}{x^2} \, dx = \int_1^n \frac{1}{x^2} \, dx + \int_2^n \frac{1}{x^2} \, dx + \int_3^n \frac{1}{x^2} \, dx$$

$$+ \int_4^n \frac{1}{x^2} \, dx + \cdots + \int_{n-1}^n \frac{1}{x^2} \, dx.$$

In this sum the interval $[1, 2)$ is covered just once, the interval $[2, 3)$ is covered twice, the interval $[3, 4)$ is covered three times, ... the interval $[n - 1, n)$ is covered $n - 1$ times, and that is precisely what Equation (12.1) is saying. The two are the same. Therefore,

$$\int_1^n \frac{\lfloor x \rfloor}{x^2} \, dx = \sum_{1 \leqslant r \leqslant n} \int_r^n \frac{1}{x^2} \, dx$$

$$= \sum_{r=1}^n \left(\frac{1}{r} - \frac{1}{n} \right) = H_n - n \times \frac{1}{n} = H_n - 1.$$

So,

$$H_n = 1 + \int_1^n \frac{\lfloor x \rfloor}{x^2} \, dx = 1 + \int_1^n \frac{x - \{x\}}{x^2} \, dx$$

$$= 1 + \int_1^n \frac{x}{x^2} - \frac{\{x\}}{x^2} \, dx = 1 + \int_1^n \frac{1}{x} - \frac{\{x\}}{x^2} \, dx$$

$$= 1 + \ln n - \int_1^n \frac{\{x\}}{x^2} \, dx.$$

This means that

$$\gamma = \lim_{n \to \infty} (H_n - \ln n) = 1 - \int_1^\infty \frac{\{x\}}{x^2} \, dx \quad \text{and} \quad \int_1^\infty \frac{\{x\}}{x^2} \, dx = 1 - \gamma,$$

as required.

With Euler (and us) failing to identify γ in terms of the logarithm of an important number, we mentioned on p. 52 that he provided a number of formulae for its evaluation, one of which was

$$\sum_{i=2}^\infty \frac{1}{i}(\zeta(i) - 1) = 1 - \gamma,$$

which he used to calculate the value to five decimal places. We will now derive his expression:

$$\gamma = \lim_{n \to \infty} \left(\sum_{r=1}^n \frac{1}{r} - \ln n \right) = \lim_{n \to \infty} \left(\sum_{r=1}^n \frac{1}{r} - \sum_{r=2}^n \ln \left(\frac{r}{r-1} \right) \right)$$

$$= 1 + \lim_{n \to \infty} \left(\sum_{r=2}^n \left(\frac{1}{r} - \ln \left(\frac{r}{r-1} \right) \right) \right) = 1 + \sum_{r=2}^\infty \left(\frac{1}{r} + \ln \left(\frac{r-1}{r} \right) \right)$$

$$= 1 + \sum_{r=2}^\infty \left(\frac{1}{r} + \ln \left(1 - \frac{1}{r} \right) \right) = 1 + \sum_{r=2}^\infty \left(\frac{1}{r} - \sum_{i=1}^\infty \frac{1}{ir^i} \right)$$

$$= 1 - \sum_{r=2}^\infty \left(\sum_{i=2}^\infty \frac{1}{ir^i} \right) = 1 - \sum_{i=2}^\infty \left(\sum_{r=2}^\infty \frac{1}{ir^i} \right)$$

$$= 1 - \sum_{i=2}^\infty \left(\frac{1}{i} \sum_{r=2}^\infty \frac{1}{r^i} \right) = 1 - \sum_{i=2}^\infty \frac{1}{i}(\zeta(i) - 1)$$

and the result clearly follows.

Yet again we have used the expansion of $\ln(1 - x)$ to eliminate the logarithm.

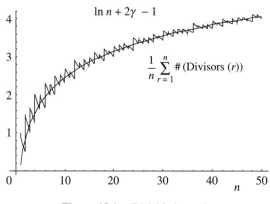

Figure 12.1. Dirichlet's result.

Following Euler's path a little further, we need 15 terms of the series to achieve five decimal places of accuracy, and so

$$\gamma \approx 1 - \tfrac{1}{2}(\zeta(2) - 1) - \tfrac{1}{3}(\zeta(3) - 1) - \tfrac{1}{4}(\zeta(4) - 1)$$
$$\qquad - \tfrac{1}{5}(\zeta(5) - 1) - \cdots - \tfrac{1}{15}(\zeta(15) - 1)$$
$$\quad = 1 - (0.322\,467 + 0.067\,352\,3 + 0.020\,580\,8 + \cdots + 0.000\,002\,039\,22)$$
$$\quad = 0.577\,217\ldots,$$

which is perfectly easy to calculate with a modern computer running state-of-the-art software...

12.3 In Number Theory

Although γ's appearance in number theory is no matter for surprise, the manner of its appearance can be puzzling. We will list just a few ways in which it emerges.

- In 1838 Lejeune Dirichlet (1805–1859) proved that

$$\frac{1}{n} \sum_{r=1}^{n} \#(\text{divisors}(r)),$$

the average number of divisors of all integers from 1 to n, approaches $\ln n + 2\gamma - 1$ as n increases (see Figure 12.1).

Further along the line,

$$\frac{1}{1000} \sum_{r=1}^{1000} \#(\text{divisors}(r))$$

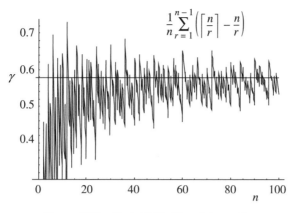

Figure 12.2. De la Vallèe Poussin's result.

evaluates to 7.069, and $\ln 1000 + 2\gamma - 1 = 7.062\,19\ldots$.

- Equally baffling, in 1898 Charles de la Vallèe Poussin (1866–1962) (more of him later) proved that if we divide an integer n by all integers less than it and average the deficits of each quotient to the integer above it, the answer approaches γ as $n \to \infty$. This time the calculation is

$$\frac{1}{n} \sum_{r=1}^{n-1} \left(\left\lceil \frac{n}{r} \right\rceil - \frac{n}{r} \right),$$

with the graph shown Figure 12.2.

And, again, further down the line we have

$$\frac{1}{10\,000} \sum_{r=1}^{9999} \left(\left\lceil \frac{10\,000}{r} \right\rceil - \frac{10\,000}{r} \right),$$

which evaluates to $0.577\,216\ldots$.

Incredibly, the result remains true if the divisors are those in any arithmetic sequence or if they are only the prime divisors.

- γ also appears (rather messily) in three standard asymptotic measures of the efficiency of the Euclidean Algorithm. In each case it appears because of the implicit appearance of the Glaisher–Kinkelin constant, mentioned on p. 88, and the explicit appearance of Porter's constant, which is the impressive

$$\frac{6\ln 2}{\pi^2} \left(3\ln 2 + 4\gamma - \frac{24}{\pi^2}\zeta'(2) - 2 \right) = 1.467\,07\ldots.$$

113

- On p. 34 we mentioned Euler's result regarding the possibility of representing an integer as the sum of two squares. If it is possible to do so, asymptotic estimates of the number of possible ways involve the Sierpinski constant $(2.584\,981\,7\ldots)$, which itself involves γ—and which is rather too cumbersome to define!

- To understand the last example it is necessary to appreciate a convergence that might be extracted from a divergent sequence. If an infinite sequence $\{a_n\}$ converges to a limit l, any infinite subsequence will converge to that same limit. That is perfectly reasonable. Now suppose that the sequence does not converge and that we consider the set of limits L of all infinite subsequence that do converge (a technical result known as the Bolzano–Weierstrass Theorem ensures that there is at least one such), then L has a maximum and a minimum. If we write the maximum as l^- and the minimum as l_-, these are called superior and inferior limits, respectively, and they are written as

$$l^- = \limsup_{n\to\infty} a_n \quad \text{and} \quad l_- = \liminf_{n\to\infty} a_n.$$

For example, the oscillating sequence $-1, 1, -1, 1, -1, 1, \ldots$ clearly does not converge but has the two convergent subsequences $1, 1, 1, 1, 1, 1, \ldots$ and $-1, -1, -1, -1, \ldots$ with limits of 1 and -1, of course. This means that

$$\limsup_{n\to\infty} a_n = 1 \quad \text{and} \quad \liminf_{n\to\infty} a_n = -1.$$

A little more subtly,

$$\frac{1}{2}, \frac{1}{3}, \frac{2}{3}, \frac{1}{4}, \frac{2}{4}, \frac{3}{4}, \frac{1}{5}, \frac{2}{5}, \frac{3}{5}, \frac{4}{5}, \frac{1}{6}, \ldots$$

does not converge but the subsequences $\frac{1}{2}, \frac{1}{3}, \frac{1}{4}, \frac{1}{5}, \ldots$ and $\frac{1}{2}, \frac{2}{3}, \frac{3}{4}, \frac{4}{5}, \ldots$ converge to 0 and 1, respectively, and so

$$\limsup_{n\to\infty} a_n = 1 \quad \text{and} \quad \liminf_{n\to\infty} a_n = 0.$$

With these ideas in place, we can at once mention an important idea in the study of primes, list a truly impressive-looking formula, mention Erdos once again, give another example of γ appearing and reveal a (typically poor) mathematical joke. The length of the interval between consecutive primes, $p_{n+1} - p_n$, is of clear importance and one of the consequences of the Prime Number Theorem (which we are inexorably approaching) is that, on average, $p_{n+1} - p_n$ is about $\ln p_n$. That said, the average in no way typifies the sequence's behaviour, as $p_{n+1} - p_n$ oscillates wildly and is very much a contender for $\limsup_{n\to\infty}$ and $\liminf_{n\to\infty}$ investigation. The latter is the more problematic, as it is not even known if

$$\liminf_{n\to\infty}(p_{n+1} - p_n) < \infty,$$

although Erdos (and others) have made some progress with this. It is the $\limsup_{n\to\infty}$ that provides our stupendous formula, which is a 1990 result of Maier and Pommerance, following a 1935 result of Erdos and also a number of others in between,

$$\limsup_{n\to\infty} \frac{(p_{n+1} - p_n)(\log\log\log p_n)^2}{(\log p_n)(\log\log p_n)(\log\log\log\log p_n)} \geqslant \frac{4e\gamma}{c},$$

where $c = 3 + e^{-c}$. Any comment would seem superfluous. It is the natural logarithm, but to use ln would be to deny the opportunity to mention the joke: what noise does a drowning Analytic Number Theorist make? Log...log...log...log...

With this idea in place, we have finally the wildly divergent sequence generated by Euler's curiously named Totient function $\varphi(n)$ (presumably from the Latin 'tot', which means 'so much'), which is defined to be the number of positive integers not greater than n and co-prime to n. It finds extensive use in very many number-theoretic investigations. Edmund Landau (1877–1938) proved that

$$\limsup_{n\to\infty} \varphi(n) = 1$$

but that

$$\liminf_{n\to\infty} \frac{\varphi(n)\ln\ln n}{n} = e^{-\gamma}.$$

He also proved that for N large,

$$\sum_{n=1}^{N} \frac{1}{\varphi(n)} \approx A\ln N + B,$$

where A is the elegant

$$\frac{\zeta(2)\zeta(3)}{\zeta(6)}$$

and B is distinctly inelegant but its expression contains π, $\zeta(3)$ and γ.

As an example of the elegance and usefulness of the Totient function, the reader should be aware that it might help with a route to mathematical immortality in that if the Goldbach Conjecture is true (every even number greater than 2 is the sum of two primes), then for all integers n there are primes p, q such that $\varphi(p) + \varphi(q) = 2n$.

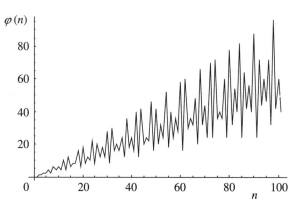

Figure 12.3. Euler's Totient function.

12.4 IN CONJECTURE

- Suppose that we toss a fair coin indefinitely and record the sequence of heads and tails. Now we choose an integer n and list all 2^n possible sequences of heads and tails. How many times will we have to toss the coin in order to see each of our sequences appear? It is known that the minimum possible number of tosses is $2^n + n - 1$ and it is conjectured that the average number approaches $2^n(\gamma + n \ln 2)$ for large n.

- A second conjecture concerns Mersenne primes, which are primes of the form $2^p - 1$, where p is prime (a natural hunting ground for big primes). It has been conjectured that if $M(x)$ is the number of primes $p \leqslant x$ for which $2^p - 1$ is prime then $M(x) \sim k \ln x$, where $k = e^{\gamma}/\sqrt{2}$. Since there are only 39 known Mersenne primes, the evidence has to be considered a touch scanty.

12.5 IN GENERALIZATION

Carl Gustav Jacobi (1804–1851) is quoted as saying, 'One should always generalize', and such a view is very much part of mathematical philosophy, but there are often several directions in which the generalization could be made. So it is with γ.

- We could move into two dimensions—but how? We will describe one way, which leads to the Masser–Gramain constant and requires a different approach to the harmonic series. Take the real line and the positive integer points on it and select the origin as a fixed reference point, then the interval [0, 1], of length 1, is the smallest interval containing the integer 1; the interval [0, 2], of length 2, is the smallest interval containing the integer

2, etc., and we can think of the expression

$$\lim_{n\to\infty}\left(\sum_{r=1}^{n}\frac{1}{r}-\ln n\right)$$

as

$$\lim_{n\to\infty}\left(\sum_{r=1}^{n}\frac{1}{\text{length of interval }[0,r]}-\ln n\right).$$

Having generalized 'interval', we now need to generalize 'integer'. For 300 years, Fermat's Last Theorem had been a 'primordial soup' from which vast and vastly important areas of mathematics have been developed. We have already mentioned its connection with the Bernouilli Numbers. With Andrew Wiles (born 1953) finally settling the matter, it may be that it has given its final 'birth', but in the mathematical ferment of the 19th century it brought about the development of numbers of the form $a+b\sqrt{-1}$, where a and b are rational numbers and, when both are integers, numbers called Gaussian integers (named after ...). Now we can move to two dimensions and from $\mathbb{R}$ to $\mathbb{C}$ to define the exotic

$$\delta=\lim_{n\to\infty}\left(\sum_{r=2}^{n}\frac{1}{\pi(\rho_r)^2}-\ln n\right),$$

the Masser–Gramain constant (we need to start at 2 to make the definition sensible). The denominators are the two-dimensional equivalent of interval length: the areas of circles and the ρ_r are defined by

$$\rho_r=\min\{\rho:\text{ there is a closed disc with radius }\rho\text{ containing}$$
$$\text{at least }r\text{ distinct Gaussian integers}\}.$$

Perhaps it is not surprising that the exact value of the constant isn't known!

Euler (naturally) embraced the idea of generalization and did so by considering

$$\lim_{n\to\infty}\left(\sum_{r=1}^{n}\frac{1}{r}-\ln n\right)$$

as

$$\lim_{n\to\infty}\left(\sum_{r=1}^{n}f(r)-\int_{1}^{n}f(x)\,dx\right)$$

with $f(x)=1/x$ as just one particular positive, decreasing function. From this he generalized to

$$f(x)=\frac{1}{x^\alpha},\quad\text{where }0<\alpha<1,$$

117

to produce two divergent components that combine to converge to finite sums—known as Euler's generalized constants.

With $f(x) = \ln^m x/x$, with m a positive integer, we have a final family of generalizations to which we will refer, known as Stieltjes constants γ_m, about which not so very much is known. Their definition is

$$\gamma_m = \lim_{n \to \infty} \left(\sum_{r=1}^{n} \frac{\ln^m r}{r} - \int_1^n \frac{\ln^m x}{x} \, dx \right)$$

$$= \lim_{n \to \infty} \left(\sum_{r=1}^{n} \frac{\ln^m r}{r} - \left[\frac{\ln^{m+1} x}{m+1} \right]_1^n \right)$$

$$= \lim_{n \to \infty} \left(\sum_{r=1}^{n} \frac{\ln^m r}{r} - \frac{\ln^{m+1} n}{m+1} \right).$$

Of course,

$$\gamma_0 = \lim_{n \to \infty} \left(\sum_{r=1}^{n} \frac{1}{r} - \frac{\ln n}{1} \right) = \gamma.$$

These are of particular importance because of their appearance in the series expansion of the complex form of the Zeta function (it is called the Laurent expansion, which is discussed in Appendix D). To be exact,

$$\zeta(z) = \frac{1}{z-1} + \sum_{r=0}^{\infty} \frac{(-1)^r}{r!} \gamma_r (z-1)^r.$$

There are other generalizations (including a lattice sum form of immense complexity) but we hope that by now the point is made that Euler's simple and natural original definition can lead to interesting and sometimes important extensions. To paraphrase Andrew Wiles, 'we think we will stop here'.

It's a Harmonic World

I tell them that if they will occupy themselves with the study of mathematics they will find in it the best remedy against the lusts of the flesh.

Thomas Mann (1875–1955)

We will now take a brief look at several of the ways in which H_n appears, and the pattern of numbers $1, \frac{1}{2}, \frac{1}{3}, \ldots$ forming its terms appear, in some areas of considerable diversity. The selection is by no means comprehensive and each initiative can be developed (in some cases very considerably) beyond where we leave it, but to delve deeper or to embrace more widely would engulf more pages than this book could afford. Firstly, though, we ought to address the question of the name 'harmonic'.

13.1 WAYS OF MEANS

With two numbers a and b, if one had to write down three examples of an average of two numbers a and b, it is likely that they would be (in order) their arithmetic, geometric and harmonic (or subcontrary) means, defined by

$$A = \tfrac{1}{2}(a+b), \qquad G = \sqrt{ab}, \qquad H = \frac{2}{1/a + 1/b},$$

respectively, and there is a nice order to them and relationship between them.

The Babylonian identity $ab = \frac{1}{4}((a+b)^2 - (a-b)^2)$, which we mentioned on p. 1, can be rewritten as

$$\left(\frac{a+b}{2}\right)^2 = ab + \left(\frac{a-b}{2}\right)^2,$$

which tells us that

$$\left(\frac{a+b}{2}\right)^2 \geqslant ab,$$

119

and therefore that $A \geqslant G$. It is also perfectly clear that both A and G lie between a and b. Also,

$$H = \frac{ab}{\frac{1}{2}(a+b)} = \frac{G^2}{A},$$

which gives us the pretty $G = \sqrt{AH}$ and the order $H \leqslant G \leqslant A$. Chasing inequalities easily shows that H is greater than the smaller of a and b and so all three means are nicely ordered within the interval, which is reasonable—but there are other definitions of means. Although our interest is really with H, it would be a shame to omit at least some mention of where it and the other two definitions fit in to the greater scheme of things.

The famous theorem that bears his name evidences a tiny part of what Pythagoras of Samos (*ca.* 569–*ca.* 475 B.C.) helped to bring to the world, a sentiment agreed by Bertrand Russell (1872–1970), who said,

> It is to this gentleman that we owe pure mathematics. The contemplative ideal, since it led to pure mathematics, was the source of a useful activity. This increased its prestige and gave it a success in theology, in ethics, and in philosophy.

The distinction between what Pythagoras himself discovered and what was discovered by his clandestine society is impossible to make, so secretive were they, but it is clear that he knew of the three means that we have mentioned. It is also clear that later Pythagoreans defined at least seven more as part of the following general schema.

Given two numbers a and c, define a number b to be a 'mean' of the other two, such that $a \leqslant b \leqslant c$. If this inequality holds, then $b - a$, $c - b$ and $c - a$ are all $\geqslant 0$ and the Pythagoreans investigated the idea of comparing the ratios of pairs of these differences with the (not necessarily distinct) ratios of the original numbers. For example, if we take the ratio

$$\frac{b-a}{c-b} = \frac{a}{a} = \frac{b}{b} = \frac{c}{c}$$

we will arrive at $b = \frac{1}{2}(a + c)$, or A. Alternatively, we could take

$$\frac{c-b}{b-a} = \frac{b}{a} = \frac{c}{b}$$

to get $b = \sqrt{ac}$, or G. The harmonic mean H emerges from

$$\frac{c-b}{b-a} = \frac{c}{a}.$$

Playing with the possibilities, as no doubt the Pythagoreans did, results in several novel definitions of mean, for example

$$\frac{c-b}{b-a} = \frac{a}{c}$$

reduces to the elegant symmetric mean

$$S = \frac{a^2 + c^2}{a + c},$$

whereas

$$\frac{c - b}{b - a} = \frac{a}{b}$$

produces the distinctly inelegant and unsymmetrical

$$b = \frac{c - a + \sqrt{a^2 - 2ac + 5c^2}}{2}.$$

This last definition recovers a little of its dignity by giving the mean of 1 and 2 as the Golden Ratio $\varphi = 1.618\,033\,9\ldots$. The reader may wish to investigate the other alternatives, some of which collapse, while others are as strange as the example above. All but the first three of the definitions have disappeared through the millennia, but a definition of mean which is important to this day is missing—the 'root mean square'

$$\sqrt{\frac{a^2 + c^2}{2}},$$

but the reader can check that this can be recovered as $\sqrt{2A^2 - G^2}$ or $\sqrt{AS}$.

Generalizations of the definitions of arithmetic, geometric and harmonic means to n numbers are obvious and we will have need of them in later chapters.

In fact, generalizations of the definition of means exist in the modern day, notably with

- Hölder's means, defined by $H_p(a, c) = \left[\dfrac{a^p + c^p}{2}\right]^{1/p}, p \neq 0;$

- Lehmer's means, defined by $L_p(a, c) = \dfrac{a^p + b^p}{a^{p-1} + b^{p-1}};$

- Stolarsky's means, defined by $S_p(a, c) = \left[\dfrac{a^p - c^p}{pa - pc}\right]^{1/(p-1)}, p \neq 0, 1.$

And it is easy to see that $A = H_1 = L_1 = S_2, G = \lim_{p \to 0} H_p = L_{1/2} = S_{-1}$ and $H = H_{-1} = L_0$.

13.2 GEOMETRIC HARMONY

The Pythagoreans held that 'All things consist of number', that is, positive integers or their ratios—and preferably small integers too. Integers were endowed with qualitative attributes such as gender that today belong to the world of

Table 13.1. Pythagorean solids.

	Faces f	Vertices v	Edges e
Cube	6	8	12
Tetrahedron	4	4	6
Octahedron	8	6	12
Icosahedron	20	12	30
Dodecahedron	12	20	30

Table 13.2. Harmonic polyhedra.

f	e	v
6	12	8
30	70	42
170	408	240

the numerologist rather than the mathematician, yet some concepts have carried over the millennia, for example figurate numbers (square, triangular, cubic, pyramidal, etc.). His mysticism had the five 'Pythagorean solids' known to him, the cube, tetrahedron, octahedron, icosahedron and dodecahedron, coupled with earth, fire, air, water and aether (see Table 13.1). The Pythagorean, Philolaus, is said to have called the cube 'a geometric harmony' because the numbers 6, 8 and 12 are in harmonic progression, with 8 the harmonic mean of 6 and 12 (but then so is the octahedron if the order of the numbers does not matter), which leads to a nice question (asked and answered by John Webb) about which other polyhedra are harmonic in the sense of Philolaus.

Yet another of Euler's results helps to provide the answer; the fundamentally important topological fact that for any convex polyhedron the number of vertices, faces and edges are related by $v + f = 2 + e$. Add to this the condition of 'geometric harmony' that

$$v = \frac{2}{1/e + 1/f}$$

and some of Webb's judicious algebra and we have that

$$(e - f - 1)^2 - 2(f - 1)^2 = -1,$$

which is Pell's equation. The continued-fraction approximations of $\sqrt{2}$ then yield a list of the infinite number of possibilities for 'harmonic polyhedra', which starts with the values given in Table 13.2. The first one is called a cube...

Table 13.3. Pythagorean musical scale.

Note	do	re	mi	fa	so	la	ti	do
Ratio	1:1	9:8	81:64	4:3	3:2	27:16	243:128	2:1

13.3 MUSICAL HARMONY

The sequence of 6, 8 and 12 appear again in the Pythagorean world. If a string of length 12 units emits a middle C when plucked, the same string shortened to 8 units will emit G, a perfect fifth above C, and if the string is shortened to 6 units it will emit C an octave up. The young Pythagoras is thought to have noticed such behaviour through hearing the variously concordant and discordant sounds of blacksmiths' hammers sounding together and, through his subsequent investigations with stretched strings, is credited with the discovery that musical intervals which are recognized as concordant are related by small integer ratios. More generally, a half length gives a frequency ratio of 2:1, the musical octave, a third length gives a ratio of 3:2, the musical fifth, a quarter length gives a frequency ratio 4:3, the musical fourth, a frequency ratio 5:4, the major third. That the arithmetic sequence 1, 2, 3, 4, 5 is involved only strengthened the belief in the sacred nature of number. It is easy to see that the reciprocals of any sequence of numbers in arithmetic progression are themselves in harmonic progression; the Pythagoreans knew this too, and so we arrive at the modern definition of a harmonic sequence. In the attributed words of the Pythagorean, Iamblichus:

> the harmonic mean was then called subcontrary, but which was renamed harmonic by the circle of Archytas and Hippasus, because it seemed to furnish harmonius and tuneful ratios.

As we discuss the Pythagorean contribution to musical theory we should mention the musical scale that bears his name, which was based on the view that the fifth is a particularly pleasing ratio and that the scale should be constructed from it and the 2:1 octave. So, taking 'fifths of fifths' and scaling down by 2 as much as necessary to bring it within the octave brings about Table 13.3 and the Greek musical scale of the Pythagorean school. The process will never fill the octave (there are plenty of numbers missed, not least the embarrassingly irrational $\sqrt{2}$) and it will never reach an octave exactly since no power of $\frac{3}{2}$ is a power of 2. To be so would mean that $(\frac{3}{2})^n = 2^m$ or $3^n = 2^{m+n}$, which brings in logarithms with

$$\frac{m+n}{n} = \frac{\ln 3}{\ln 2} = 0.405\,465\ldots,$$

Table 13.4. Gradus Suavitatis.

Ratio	Gradus suavitatis
1/2 (octave)	2
3/2 (fifth)	4
4/3 (fourth)	5
5/4 (major third), 5/3 (major sixth)	7
6/5 (minor third), 9/8 (major whole tone), 8/5 (minor sixth)	8
10/9 (minor whole tone), 9/5 (minor seventh), 15/8 (major seventh)	10
16/15 (diatonic semitone)	11
81/64 (pythagorean major third), 45/32 (tritone)	14

and continued fractions with $0.405\,465\ldots = [1, 1, 1, 2, 2, 3, 1, 5, \ldots]$ giving best approximations of

$$\frac{m+n}{n} = \frac{1}{1}, \frac{2}{1}, \frac{3}{2}, \frac{8}{5}, \frac{19}{12}, \frac{65}{41}, \ldots$$

Therefore, it is arithmetically optimal to build a scale based on the fifth if we use 1, 2, 5, 12, 41, ... of them to the octave (quite what they would sound like is another matter), which means that the Pythagorean scale is not in this sense optimal. Another way to look at the inexactitude is that in the scale the interval between successive notes is either a 'tone' of 9:8 or a 'minor semitone' of 256:243; unfortunately, the semitone is not quite half a tone since two of them give a frequency ratio of $(256:243)^2 \neq 9:8$. Factorize into primes and the error lies in the approximation that $2^{19} \approx 3^{12}$ or $(\frac{3}{2})^{12} \approx 2^7$ and so going up 12 fifths and then down 7 octaves brings you back to where you started—nearly. The difference is known as the 'Pythagorean comma', which will be our full stop!

Harmony was close to Euler's heart too. During the Middle Ages the *quadrivium* comprised the four mathematical 'arts'—arithmetic, music, geometry, astronomy—and constituted the higher part of knowledge, as opposed to the *trivium*, the elementary part, which comprised grammar, rhetoric and dialectic. Euler lived later, but it was understandable that a man such as he would take an interest in music, particularly as parts of his long life coincided with the lives of Bach, Handel, Haydn and Mozart. In 1731, when he was 24 years old, he wrote *An attempt at a new theory of music, exposed in all clearness according to the most well-founded principles of harmony* (although it was not published until 1739) and returned time and again to musical theory, refining and developing his thoughts. We will make no great attempt to pursue him here (this study alone occupied 263 pages) but simply mention his use of primes in trying to quantify the melodiousness, the 'degree of sweetness'—or as he called it the 'gradus suavitatis'—of sounds. The gradus suavitatis of a single note was taken to be

1 and beyond that, if the frequency ratio of two notes is $m:n$ and if the least common multiple of m and n is L, he made the definition

$$G(m, n) = 1 + \prod_{\substack{p \text{ prime} \\ p \text{ divides } L}} (p - 1),$$

with multiplicities taken into account. For example, $G(4, 3) = 1 + (2 - 1) + (2 - 1) + (3 - 1) = 5$ and, more fully, we have Table 13.4.

Now that we have the name of the series properly established, we will look at some of those places in which it appears.

13.4 SETTING RECORDS

A record is 'the best there is so far'. There are any number of examples of sequences of numbers naturally appearing wherein there is 'improvement' at some stage and beyond that another and H_n naturally appears in the analysis of them. For example, consider rainfall figures and assume that the rainfall in one year does not affect that in any subsequent year; that is, annual rainfall figures are independent random variables. The first year of recording is a record by definition. In the second year, the rainfall level could equally likely be less or more than the first year, so the expected number or record years in the first two years is $1 + \frac{1}{2}$. Continue this reasoning for a third year and we have two of the six possible orderings of the rainfall for the three years having the third year as a record and so the expected number of record years is $1 + \frac{1}{2} + \frac{1}{3}$ years. Continue this reasoning for n years, and we have that the expected number of record years is

$$1 + \frac{1}{2} + \frac{1}{3} + \cdots + \frac{1}{n} = H_n.$$

Two arbitrarily chosen examples are revealing. The Radcliffe Meteorological Station in Oxford has data for rainfall in Oxford between 1767 and 2000 and there are five record years; this is a span of 234 recorded years and $H_{234} = 6.03$. For Central Park, New York City, between 1835 and 1994 there are six record years over the 160-year period and $H_{160} = 5.65$, providing good evidence that English weather is that bit more unpredictable! An interesting implication of the surprisingly small values of H_n (for example H_{1000} and $H_{1\,000\,000}$ are 7.49 and 14.39, respectively) is that, without climactic change, record years would be very rare even over these large time spans.

The accuracy of the predictions, based on the assumption of statistical independence between readings, can be turned around to itself be a measure of that independence. In particular, and to quote Ned Glick,

> ...at a 1954 meeting of the Royal Statistical Society, F. G. Foster and A. Stuart pointed out that record low and record high annual

rainfalls in Oxford were much more rare than record breaking performances (low times or high distances) in annual track and field
competitions of the British Amateur Athletic Association. This con-
trast is not surprising: athletic recruiting and training have inten-
sified over the past century; but no one has done much about the
weather. Although athletic performances do fluctuate, there is an
average trend over decades for national competitors (and therefore winners) to run faster, jump higher or throw further; while
weather conditions over a century are more intuitively random,
without dramatic linear trend. Of course, it is possible for 100 random observations to be ordered so that the sequence has as many
as 10, 50 or 100 record highs. But detailed calculation shows that
the probability of 10 or more record highs in a 100-long random
sequence is less than 5%. Therefore, in a situation where data are
less familiar than rainfalls or race times, the mere finding of many
record highs or lows suggests that the data are not a simple random
sample; that is, an alternative hypothesis should be sought to fit
the data better. Foster and Stuart gave formal procedures using the
sum or the difference of record high and record low frequencies to
fit or to test the hypothesis of randomness. Other statisticians have
also considered such inferential procedures.

13.5 TESTING TO DESTRUCTION

Suppose that we have n wooden beams that are to be used as horizontal supports
in building projects. Naturally, we would want to know how strong they are, with
the minimum breaking strain the crucial factor. To test this breaking strain we
can imagine placing a beam on two supports, one at either end, and applying a
gradually increasing force at its centre; when the beam breaks we will record its
breaking strain. Applying this technique will assuredly give us the information
we want but at the cost of the destruction of all of the beams. We will know
what was, rather than what is, true. A less expensive and more useful approach
would be this: let the breaking strain of the rth beam be B_r for $1 \leqslant r \leqslant n$, then
we adopt the following procedure.

- Test the first beam to destruction, so that we know B_1.

- Test the second beam by gradually increasing the force to B_1 but no
  further. If it survives, we will know that $B_2 > B_1$; if it breaks, we record
  its breaking strain, B_2.

- Test the third beam by gradually increasing the force to $\min\{B_1, B_2\}$. If
  it breaks, record B_3, otherwise move on to the next beam.

Using the same reasoning as with record years, given that the strength of the beams is an independent variable, the expected number of beams broken is

$$H_n = 1 + \frac{1}{2} + \frac{1}{3} + \cdots + \frac{1}{n}.$$

So, rather than breaking all of the 1000 beams, we would expect to break about $H_{1000} \approx 7.5$ of them to establish the minimum breaking strain, no doubt to the delight of the building company. It can also be shown that the variance of the number of beams broken is $H_n - \pi^2/6$, with another appearance of $\pi^2/6$.

13.6 Crossing the Desert

We will look at the problem in its form as a puzzle of World War II and solved by N. J. Fine in 1947, although it dates back further.

You have to cross the desert by jeep. There are no sources of fuel in the desert, and you cannot carry enough fuel in a jeep in order to make the crossing in one go. You do not have time to establish fuel dumps, but you do have a large supply of jeeps and drivers, none of which you want to lose. How can you get across the desert, using the minimum amount of fuel?

We will measure the distance a jeep can travel in terms of a tankful of fuel; one jeep by itself can travel a distance of one tankful. If two jeeps set out together, they should travel for $\frac{1}{3}$ of a tankful, then Jeep 2 transfers $\frac{1}{3}$ of its tankful to Jeep 1, and returns to base on the remaining $\frac{1}{3}$ tankful. Jeep 1 is then able to travel a total of $1 + \frac{1}{3}$ tankfuls.

With three jeeps, they should stop after travelling $\frac{1}{5}$ of a tankful, then transfer $\frac{1}{5}$ of a tankful from Jeep 3 into each of Jeeps 1 and 2, which are now full. Jeep 3 now has $\frac{2}{5}$ of a tankful, Jeeps 1 and 2 now proceed as before, with Jeep 2 returning with an empty tank to Jeep 3. Between them, they have enough fuel to get back to base. Meanwhile, Jeep 1 has travelled a total of $1 + \frac{1}{3} + \frac{1}{5}$ tankfuls.

The same reasoning shows that with four jeeps you can achieve a distance of $1 + \frac{1}{3} + \frac{1}{5} + \frac{1}{7}$ tankfuls, and with n jeeps you can get a jeep across a desert that is

$$\frac{1}{3} + \frac{1}{5} + \frac{1}{7} + \cdots + \frac{1}{2n - 1}$$

tankfuls wide. The divergence of the series means that with this system of transferring fuel, we can affect the crossing of a desert of arbitrarily large size—as long as there are enough jeeps and drivers.

13.7 Shuffling Cards

A 'top in at random' shuffle is one in which the top card of a card deck of n cards is removed and inserted at random in the deck. How many times must this shuffle be repeated before we can regard the deck as 'random'?

We follow the progress of the card which is initially at the bottom of the deck. This card (label it B) stays at the bottom until another card is inserted below it. Since there are n places into which a card taken from the top can go, the chance that it will go below B is $1/n$, and therefore on average it will take n 'top in at random' shuffles before a card is placed below B. With this done, the chance that a card taken from the top and inserted at random into the deck will go in below B is $2/n$ since there are now two places below B, and the expected number of shuffles needed to get a second card below B is $n/2$ and the expected number of shuffles needed to get two cards below B is $n + n/2$. Note that at this stage the cards below B are in random order. Continuing in this way, we see that the expected number of 'top in at random' shuffles needed to get B up to the top of the deck is

$$n + \frac{n}{2} + \frac{n}{3} + \frac{n}{4} + \cdots + \frac{n}{n-1} = n\left(1 + \frac{1}{2} + \frac{1}{3} + \frac{1}{4} + \cdots + \frac{1}{n-1}\right).$$

At this stage the cards below B are in random order, and just one more shuffle, which puts B at random into the deck, is needed to randomize the deck. The total number of shuffles needed is therefore

$$n + \frac{n}{2} + \frac{n}{3} + \frac{n}{4} + \cdots + \frac{n}{n-1} + 1$$
$$= n\left(1 + \frac{1}{2} + \frac{1}{3} + \frac{1}{4} + \cdots + \frac{1}{n-1} + \frac{1}{n}\right)$$
$$= nH_n.$$

For an ordinary bridge deck of 52 cards this makes it about 230 'shuffles'.

13.8 QUICKSORT

Of the many different algorithms that have been devised to sort an array of data, Quicksort (devised by C. A. R. Hoare) is favoured more than most because the time that it takes to perform a sort is usually comparatively short.

The general idea of Quicksort is that an item of the array, called the pivot point, is selected and the array divided into two, with all items with a value less than the pivot point moved to or remain on its left and all items with a value more than the pivot point are moved to or remain on its right. The process is then continually repeated in each sub-array until the data are sorted, which occurs when the length of each sub-array is 1; no effort is made to arrange the data in each sub-array. To look at the mathematics involved, write T_n for the average time for the algorithm to sort a list of n items arranged in some unknown order. Suppose that the rth element of the list is chosen as the initial pivot point (which we assume takes 1 unit of comparison time), then we need $n - 1$ comparisons to divide the data into the two partitions plus the 1 and so

$$T_n = n + T_{r-1} + T_{n-r}, \quad r = 1, 2 \ldots, n \quad \text{with } T_0 = 0.$$

We can eliminate r by summing over it to give

$$\sum_{r=1}^{n} T_n = \sum_{r=1}^{n} n + \sum_{r=1}^{n} (T_{r-1} + T_{n-r}),$$

$$nT_n = n^2 + \sum_{r=1}^{n} T_{r-1} + \sum_{r=1}^{n} T_{n-r} = n^2 + 2\sum_{r=0}^{n-1} T_r,$$

$$\therefore \quad T_n = n + \frac{2}{n} \sum_{r=0}^{n-1} T_r.$$

This makes

$$nT_n - (n-1)T_{n-1} = n\left\{ n + \frac{2}{n} \sum_{r=0}^{n-1} T_r \right\} - (n-1)\left\{ n - 1 + \frac{2}{n-1} \sum_{r=0}^{n-2} T_r \right\}$$

$$= n^2 - (n-1)^2 + 2\sum_{r=0}^{n-1} T_r - 2\sum_{r=0}^{n-2} T_r = 2n - 1 + 2T_{n-1}$$

and so we get

$$nT_n = 2n - 1 + 2T_{n-1} + (n-1)T_{n-1} = (n+1)T_{n-1} + 2n - 1, \quad n = 1, 2, \ldots$$

with $T_0 = 0$.

A magical leap avoids the world of recurrence relations and we state the solution

$$T_n = 2(n+1) \sum_{r=1}^{n+1} \frac{1}{r} - 3n - 2, \quad \text{for } n \geqslant 1,$$

which we can check. If it is, then

$$(n+1)T_{n-1} + 2n - 1 = (n+1) \times \left(2n \sum_{r=1}^{n} \frac{1}{r} - 3(n-1) - 2 \right) + 2n - 1$$

$$= 2n(n+1) \sum_{r=1}^{n} \frac{1}{r} + (n+1)(-3n+1) + 2n - 1$$

$$= 2n(n+1) \sum_{r=1}^{n} \frac{1}{r} - 3n^2 - 2n + 1 + 2n - 1$$

$$= 2n(n+1) \sum_{r=1}^{n+1} \frac{1}{r} - 2n - 3n^2 = nT_n.$$

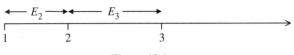

Figure 13.1.

And we are done, since we also have

$$T_0 = 2(0 + 1) \sum_{r=1}^{1} \frac{1}{r} - 3 \times 0 - 2 = 0.$$

Using this we can give a measure of the efficiency of Quicksort by estimating H_n by the natural logarithm and replacing n by $n + 1$ for large n:

$$T_n = 2(n + 1) \sum_{r=1}^{n+1} \frac{1}{r} - 3n - 2 \approx 2(n + 1)\left(\ln(n + 1) + \gamma - \frac{3n + 2}{2(n + 1)}\right)$$

$$= O(n \ln n).$$

This compares pretty favourably with alternatives like the simple Bubble sort, where the same average is about $O(n^2)$; of course, worst-case scenarios can happen, which force the $n \ln n$ towards n^2.

13.9 COLLECTING A COMPLETE SET

There are many occasions when, as a marketing ploy, sets of objects are distributed among products to encourage sales, particularly among children. We will model the situation with packets of breakfast cereal and suppose that there are n distinct toys distributed randomly (which is a big assumption), one to each box, and among an unlimited number of boxes. The question is: what is the expected number of boxes that must be bought for the child to collect the whole set of toys?

First, we need a preliminary result. The infinite geometric series,

$$1 + x + x^2 + x^3 + \cdots = \frac{1}{1 - x} \quad \text{for } |x| < 1,$$

which we have used several times before, can legitimately be differentiated with respect to x to give

$$1 + 2x + 3x^2 + 4x^3 + \cdots = \frac{1}{(1 - x)^2},$$

with the same range of convergence. Now to the problem at hand.

Let E_r be the expected number of boxes to be opened to collect the rth new toy. Pictorially, see Figure 13.1. Since the first box must yield a new toy, $E_1 = 1$.

Then

$$
\begin{aligned}
E_2 &= 1\frac{n-1}{n} + 2\frac{1}{n}\frac{n-1}{n} + 3\left(\frac{1}{n}\right)^2\frac{n-1}{n} + 4\left(\frac{1}{n}\right)^3\frac{n-1}{n} + \cdots \\
&= \frac{n-1}{n}\left(1 + 2\left(\frac{1}{n}\right) + 3\left(\frac{1}{n}\right)^2 + 4\left(\frac{1}{n}\right)^3 + \cdots\right).
\end{aligned}
$$

Putting $x = 1/n$ in the above result then gives

$$
E_2 = \frac{n-1}{n}\frac{1}{(1-1/n)^2} = \frac{n}{n-1}.
$$

Continuing this argument,

$$
\begin{aligned}
E_3 &= 1\frac{n-2}{n} + 2\frac{2}{n}\frac{n-2}{n} + 3\left(\frac{2}{n}\right)^2\frac{n-2}{n} + 4\left(\frac{2}{n}\right)^3\frac{n-2}{n} + \cdots \\
&= \frac{n-2}{n}\left(1 + 2\left(\frac{2}{n}\right) + 3\left(\frac{2}{n}\right)^2 + 4\left(\frac{2}{n}\right)^3 + \cdots\right) \\
&= \frac{n-2}{n}\frac{1}{(1-2/n)^2} = \frac{n}{n-2}.
\end{aligned}
$$

And so the expected number of cereal boxes that must be bought to collect the whole set of toys is

$$
\begin{aligned}
T_n &= E_1 + E_2 + E_3 + \cdots + E_n \\
&= \sum_{r=1}^{n}\frac{n}{n-r+1} = n\sum_{r=1}^{n}\frac{1}{r} = nH_n.
\end{aligned}
$$

A non-random distribution would, of course, increase this number. The reader could model this by, for example, throwing a fair six-sided die until all six numbers have shown uppermost, in which case, $n = 6$ and $T_6 = 14.7$. Using an ordinary bridge deck of cards and cutting until all cards have shown would require $n = 52$ and a lot more patience: $T_{52} \approx 205$.

13.10 A PUTNAM PRIZE QUESTION

The William Lowell Putnam Mathematical Competition is an annual contest for college students in America, established in 1938 in memory of its namesake. It awards cash prizes to both individuals and teams. Problem B5 of the 1992 Putnam competition involved determinants and was the following. Is $\Delta_n/n!$

bounded, where

$$\Delta_n = \begin{vmatrix} 3 & 1 & 1 & 1 & \cdots & 1 \\ 1 & 4 & 1 & 1 & \cdots & 1 \\ 1 & 1 & 5 & 1 & \cdots & 1 \\ 1 & 1 & 1 & 6 & \cdots & 1 \\ \cdot & \cdot & \cdot & \cdot & \cdot & \cdot \\ 1 & 1 & 1 & 1 & \cdots & n+1 \end{vmatrix}.$$

We show the essential steps of an elegant solution of this problem and leave it to the interested reader to provide the details of the uses of determinants!

$$\Delta_n = \begin{vmatrix} 3 & 1 & 1 & 1 & \cdots & 1 & 1 \\ 1 & 4 & 1 & 1 & \cdots & 1 & 1 \\ 1 & 1 & 5 & 1 & \cdots & 1 & 1 \\ 1 & 1 & 1 & 6 & \cdots & 1 & 1 \\ \cdot & \cdot & \cdot & \cdot & \cdot & 1 & 1 \\ 1 & 1 & 1 & 1 & \cdots & n+1 & 1 \\ 0 & 0 & 0 & 0 & \cdots & 0 & 1 \end{vmatrix}$$

$$= \begin{vmatrix} 2 & 0 & 0 & 0 & \cdots & 0 & 1 \\ 0 & 3 & 0 & 0 & \cdots & 0 & 1 \\ 0 & 0 & 4 & 0 & \cdots & 0 & 1 \\ 0 & 0 & 0 & 5 & \cdots & 0 & 1 \\ \cdot & \cdot & \cdot & \cdot & \cdot & 0 & 1 \\ 0 & 0 & 0 & 0 & \cdots & n & 1 \\ -1 & -1 & -1 & -1 & \cdots & -1 & 1 \end{vmatrix}$$

$$= \begin{vmatrix} 2 & 0 & 0 & 0 & \cdots & 0 & 1 \\ 0 & 3 & 0 & 0 & \cdots & 0 & 1 \\ 0 & 0 & 4 & 0 & \cdots & 0 & 1 \\ 0 & 0 & 0 & 5 & \cdots & 0 & 1 \\ \cdot & \cdot & \cdot & \cdot & \cdot & 0 & 1 \\ 0 & 0 & 0 & 0 & \cdots & n & 1 \\ 0 & 0 & 0 & 0 & \cdots & 0 & H_n \end{vmatrix} = n!H_n.$$

So, $\Delta_n / n! = H_n$ and since H_n diverges, the answer is that the determinant is unbounded.

13.11 MAXIMUM POSSIBLE OVERHANG

If a stack of playing cards (for example) is placed on the edge of a table and made to overhang as in Figure 13.2, we can ask the question, What is the biggest overhang possible? Suppose that the cards are 2 units wide. Clearly, we

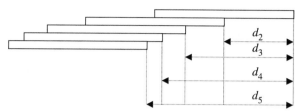

Figure 13.2. Overhanging cards.

get maximum overhang from one card to the next if the upper card is displaced so that its centre of gravity is just above the edge of the one it is on top of. Let d_r be the distance from the right edge of the top card to the same edge of the rth card from the top. Then $d_1 = 0$ and if d_{r+1} is to be the centre of gravity of the first r cards,

$$d_{r+1} = \frac{(d_1 + 1) + (d_2 + 1) + (d_3 + 1) + \cdots + (d_r + 1)}{r}, \quad 1 \leqslant r \leqslant n.$$

Hence

$$r d_{r+1} = r + d_1 + d_2 + \cdots + d_{r-1} + d_r, \quad r \geqslant 0,$$

and

$$(r - 1)d_r = r - 1 + d_1 + d_2 + \cdots + d_{r-1}, \quad r \geqslant 1.$$

Subtracting gives

$$r d_{r+1} - (r - 1)d_r = 1 + d_r, \quad r \geqslant 1.$$

And, therefore,

$$d_{r+1} = d_r + 1/r, \quad r \geqslant 1,$$

the second formula defining the harmonic series, and so $d_{r+1} = H_r$, and setting $r = n$ gives H_n as the total overhang, and again the divergence of H_n means that theoretically the overhang can be as large as we please.

13.12 WORM ON A BAND

This intriguing problem seems to have been invented by Denys Wilquin in 1972. A (mathematical) worm starts at the end of a (mathematical) rubber band of initial length 1 m. The worm crawls at a constant 1 cm min^{-1} and at the end of each minute the band instantly stretches by 1 m. So, just after 1 min of crawling the worm is 1 cm from the start and 99 cm from the end, but the band then instantly stretches by 1 m with the worm stationary relative to it, and as it is 1% from the start and 99% from the end it is 2 cm from the start and 198 cm from the end. Just at the end of the second minute the worm is 3 cm from the start and 197 cm from the end, that is, $(1 + \frac{1}{2})\% = 1.5\%$ from the start and 98.5% from the end; the band stretches again and becomes 3 m long; the worm is, therefore, 4.5 cm from the start and 295.5 cm from the end. Just at the end of the third minute the worm is 5.5 cm from the start and 194.5 cm from the end, that is, $(1 + \frac{1}{2} + \frac{1}{3}) = 1\frac{5}{6}\%$ from the start and $98\frac{1}{6}\%$ from the end. And so the process continues. The question is, does the worm ever reach the end? The answer relies critically on the fact that when the rubber band stretches the percentage of it that the worm has crawled along remains constant; therefore, he crawls $\frac{1}{100}$th of its length in the first minute, $\frac{1}{200}$th in the second minute, $\frac{1}{300}$th in the third, etc. So, after n minutes the fraction of the band that he has crawled is

$$\frac{1}{100}\left(\frac{1}{1} + \frac{1}{2} + \frac{1}{3} + \cdots + \frac{1}{n}\right) = \frac{H_n}{100}.$$

Again, using the logarithmic estimate for H_n we have that $H_n = 100$ when $\ln n + \gamma \approx 100$, which is when, $n \approx e^{100-\gamma}$ minutes. Our tireless worm will need longer than the estimated life of the Universe to complete his journey.

13.13 Optimal Choice

This final, surprising appearance of the harmonic series is remarkable for its particularly counterintuitive nature and appears in many forms: picking a secretary, a suitor, a car, a restaurant, etc. The common ground is that there is a list from which a single choice has to be made, the list is randomly ordered and there is a single best choice—and we would like to make it. We could, of course, appraise each candidate and so guarantee success, or at the other end of the spectrum we could be lazy and simply pick one at random; if there are n choices in total, the chances of success in picking the best would then be 1 and $1/n$, respectively. Is there an optimal strategy that fits somewhere in between, making us work a bit—but not too much? The answer is 'yes', and a very elegant 'yes' too. That strategy is to *reject the first r candidates on the list and then choose the first candidate better than the best reject.*

Why is this sensible and when is it optimal? What value does r have? Suppose that the best candidate is B, then we will fail if B is among the first r candidates and since all subsequent candidates will be compared to B we will inevitably have to choose the lucky nth candidate, otherwise we have a chance of success—but what chance? The answer depends on where B is among the remaining choices and we need to deal with each possibility separately.

If B happens to be in the $(r + 1)$th position, we will choose it for certain; this happens with probability $1/n$. Now suppose that B is in the $(r + 2)$th position, then if the occupant of the $(r + 1)$th position is the best yet we will fail in our goal by choosing it, otherwise we will choose B nonetheless. This means that we will choose B if the best yet among the first $r + 1$ choices lies among the first r of them; this occurs with probability $r/(r + 1)$. We do need B to be in the $(r + 2)$th position, and this happens with that same probability of $1/n$, so the total probability of success in this case is

$$\frac{1}{n} \times \frac{r}{r + 1}.$$

Now the process continues, supposing that B is in the $(r + 3)$th, $(r + 4)$th, ..., nth position, giving the probabilities of success as

$$\frac{1}{n} \times \frac{r}{r + 2}, \qquad \frac{1}{n} \times \frac{r}{r + 3}, \qquad \ldots, \qquad \frac{1}{n} \times \frac{r}{n - 1}.$$

The total probability of success (that is, of choosing B) using this strategy is then

$$P(n, r) = \frac{1}{n}\left(1 + \frac{r}{r + 1} + \frac{r}{r + 2} + \frac{r}{r + 3} + \cdots + \frac{r}{n - 1}\right).$$

For any given n it is this probability that we wish to maximize as r varies from 0 to $n - 1$ and the harmonic series is evidently making another appearance. In terms of it we have

$$P(n, r) = \frac{1}{n}\{1 + r(H_{n-1} - H_r)\}.$$

This is easily computed for small values of n, for example, $n = 5, 10, 100, 1000$, and the behaviour is shown in Figure 13.3.

The points have been joined to emphasize the behaviour. We can clearly see a trend appearing, with the maximum value of $P(n, r)$ decreasing from just over 0.4 to something under it and achieved at a value of r slightly more than a third of n. Table 13.5 gives the maximum probabilities and the values of r at which they are achieved for the first few and several larger values of n.

From this we can see that the strategy results in a probability of success (that is, of choosing B) of at least 37% no matter how large n is; it may not be certainty but it is a great deal better that the diminishing $1/n$ of the random guess.

The full analysis of the problem again has us approximating the H_n by the natural logarithm for large n (and therefore large enough r)

$$P(n, r) \approx \frac{1}{n}\{1 + r([\ln(n - 1) + \gamma] - [\ln r + \gamma])\} = \frac{1}{n}\left\{1 + r \ln \frac{(n - 1)}{r}\right\}.$$

If we treat r as a continuous variable, we can use calculus to find the approximate coordinates of the maximum that we have seen in the plots,

$$\frac{dP(n, r)}{dr} = \frac{1}{n}\left\{\ln \frac{n - 1}{r} - r \times \frac{1}{r}\right\} = \frac{1}{n}\left\{\ln \frac{n - 1}{r} - 1\right\},$$

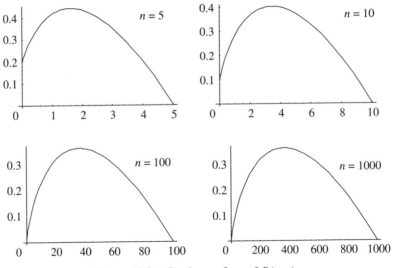

Figure 13.3. Continuous form of $P(n, r)$.

Table 13.5. Optimal choice table of values.

n	Opt. r	Max. $P(n, r)$
1	0	1.000
2	0/1	0.500
3	1	0.500
4	1	0.458
5	2	0.433
6	2	0.428
7	2	0.414
8	3	0.410
9	3	0.406
10	3	0.399
20	7	0.384
50	18	0.374
100	37	0.371 043
200	73	0.369 461
300	110	0.369 352
400	147	0.368 671
500	184	0.368512
1 000	368	0.368 195
5 000	1839	0.367 942
10 000	3678	0.367 911

Table 13.6. Another optimal choice table of values.

n	Opt. r	Max. $P(n, r)$
2	1	0.500
3	1	0.500
8	3	0.409 8
11	4	0.398 4
19	7	0.385 0
87	32	0.371 5
106	39	0.370 9
193	71	0.369 5
1264	465	0.368 13
1457	536	0.368 10

which means that for any stationary points, $\ln(n-1)/r = 1$ and so $(n-1)/r = e$ and there isn't much lost in saying $n/r = e$. So, the optimal r is about n/e and gives the maximum $P(n, r)$ approximately as

$$\frac{1}{n}\left\{1 + r\ln\frac{(n-1)}{r}\right\} \approx \frac{r}{n}\ln\frac{n}{r} \approx \frac{1}{e} \approx 0.37,$$

the base of Napier's logarithms.

A little intrigue remains, though. The continued-fraction representation of $1/e$ is just that of e shifted one place and so it is

$$1/e = [0; 2, 1, 2, 1, 1, 4, 1, 1, 6, 1, 1, 8, 1, 1, 10, 1, 1, 12, \dots].$$

This means that the first few convergents are $\frac{1}{2}, \frac{1}{3}, \frac{3}{8}, \frac{4}{11}, \frac{7}{19}, \frac{32}{87}, \frac{39}{106}, \frac{71}{193}, \frac{465}{1264}, \frac{536}{1457}, \dots$.

Table 13.5 lists some values of n and the corresponding optimal r, together with the value of $P(n, r)$. Another selection of values of n yields the equivalent Table 13.6.

The selection of the n is hardly arbitrary: they are of course the denominators of the convergents of $1/e$, and the optimum r are nothing other than the corresponding numerators. A bigger test is $n = 14\,665\,106$—the denominator of the 20th convergent of $1/e$; the numerator is $5\,394\,991$—and guess what the optimum r is? Correct. It is reasonable, but why is it true?

A peculiar feature of the procedure is that every candidate can be told the outcome of the interview immediately at its end—if they actually get an interview, that is! If we wish to sacrifice this feature, we can look at things slightly differently. Suppose that we replace the verb 'reject' with the alternative verb 'reserve', then, if the best candidate is within the first r interviewed, we will inevitably continue to interview to the last candidate, but having done that we would choose that best candidate anyway from our initial reserves. Of course,

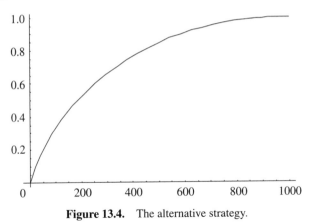

Figure 13.4. The alternative strategy.

in a sense the procedure has failed because we will have enjoyed no saving of effort but we can ask, for example, what the value of r should be to ensure that the odds of success are just in our favour. Now, we have

$$P(n, r) = \frac{1}{n}\{1 + r(H_{n-1} - H_r)\} + \frac{r}{n},$$

with the additional term the probability that the best candidate is within the first r interviewed. Even with the logarithmic approximation, finding r so that $P(n, r) = \frac{1}{2}$ is not capable of analysis but we can see what is happening with, for example, $n = 1000$.

The function is maximizing to 1 at $r = 1000$ (unsurprisingly), but we are interested in where it equals $\frac{1}{2}$ and a bit of computation reveals that this is achieved when $r = 186$ and continuing to higher n indicates that we have an asymptotic form $r/n \approx 0.186\,682\,2\ldots$—whatever that is. In summary, if we apply the procedure, having automatically interviewed a bit under 20% of the candidates we have an even chance of picking the best of them!

It's a Logarithmic World

How can it be that mathematics, being after all a product of human thought independent of experience, is so admirably adapted to the objects of reality?

Albert Einstein (1879–1955)

As we have mentioned in the Introduction, the reader of this book will need little convincing that logarithms appear with great frequency in mathematics and its applications, particularly with so many differential equations involving them or exponentials in their solution. Power laws abound in nature: Kepler's third law, the universal law of gravitation, Boyle's Law, etc. A browse through any science book will yield any number of examples, and where there is a power law, there is a linearizing logarithm, as Kepler may have experienced. The intensity of earthquakes is measured on the logarithmic Richter scale, fractal dimension is defined in terms of logarithms, distance in the Poincaré model of hyperbolic geometry is logarithmic, and so the list continues. The final two chapters are devoted to one particular and major use of them as a measure of the number of primes below any number. Here we will look at three other examples of them forcing themselves into the solution of a problem. They can hardly be representative, but each has a novel appeal and each has been developed into important ideas.

14.1 A MEASURE OF UNCERTAINTY

A dictionary definition of 'entropy' is 'a measure of the disorder of a system'. The word is famously associated with the Second Law of Thermodynamics, but in 1948 it found use in the hands of the American scientific genius Claude Shannon (1916–2001), the 'father of the information age', on whose theories rests the ideas of modern digital communication. A delicious eccentric, his house was home to five pianos and 30 other instruments, chess-playing machines (including one that moved the pieces with a three-fingered arm, beeped and made wry comments), rocket-powered Frisbees, motorized Pogo sticks, a mind-reading machine, a mechanical mouse that could navigate a maze and a device that could solve Rubik's Cube. His love of juggling led to the invention of a machine with

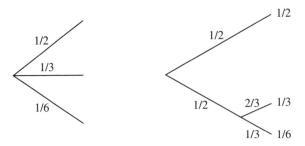

Figure 14.1. Claude Shannon's Fig. 6.

soft beanbag hands that juggled steel balls and also a tiny stage on which three clowns juggle 11 rings, 7 balls, and 5 clubs, all driven by an invisible mechanism of clockwork and rods. His love of the unicycle led to him using one as transport along the corridors of the Bell Laboratories, where he worked for many years. His love of both led to the design of a unicycle with asymmetric gearing so that he could more easily juggle... as he unicycled along.

As an employee at the Bell Telephone Company, he was naturally interested in problems that arose from communication in all its forms, an interest which led to his influential 1948 paper, later to appear in book form as *The Mathematical Theory of Communication*, co-authored with the mathematician Warren Weaver. The ideas were embraced, made rigorous and expanded by Alexandre Khinchin (whose work on continued fractions we will touch on later in this chapter) in two important papers, to appear in English in 1959 as the book *Mathematical Foundations of Information Theory*. From there, the subject has blossomed into a critically important area of modern applied mathematics. It is from the Shannon–Weaver book that our first example is culled, as they quantify the concept of the disorder in a communication system, phrasing the idea in terms of probabilities. We will not move further to see him develop the initiative into a series of seminal results, crucial to modern communication systems, although the reader may well wish to consult either book to take the study further; both are currently available. How can uncertainty be measured and how do logarithms naturally appear as a measure? We let Claude Shannon tell us (also see Figure 14.1):

6 Choice, uncertainty and entropy

We have represented a discrete information source as a Markoff process. Can we define a quantity which will measure, in some sense, how much information is 'produced' by such a process, or better, at what rate information is produced?

Suppose that we have a set of possible events whose probabilities of occurrence are $p_1, p_2, \ldots, p_n$, These probabilities are known but that is all we know concerning which event will occur. Can we

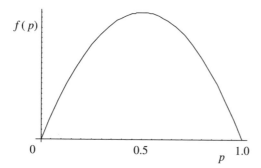

Figure 14.2. A reasonable 2-state entropy graph.

find a measure of how much 'choice' is involved in the selection of the event or of how uncertain we are of the outcome?

If there is such a measure, say $H(p_1, p_2, \ldots, p_n)$, it is reasonable to require of it the following properties:

1. H should be continuous in the p_i.

2. If all the p_i are equal, $p_i = 1/n$, then H should be a monotonically increasing function of n. With equally likely events there is more choice, or uncertainty, when there are more possible events.

3. If a choice be broken down into two successive choices, the original H should be the weighted sum of the individual values of H. The meaning of this is illustrated in Fig. 6. At the left we have three possibilities $p_1 = \frac{1}{2}$, $p_2 = \frac{1}{3}$, $p_3 = \frac{1}{6}$. On the right we first choose between two possibilities, each with probability $\frac{1}{2}$, and if the second occurs make another choice with probabilities $\frac{2}{3}$, $\frac{1}{3}$. The final results have the same probabilities as before. We require, in this special case, that

$$H(\tfrac{1}{2}, \tfrac{1}{3}, \tfrac{1}{6}) = H(\tfrac{1}{2}, \tfrac{1}{2}) + \tfrac{1}{2} H(\tfrac{2}{3}, \tfrac{1}{3}).$$

The coefficient $\frac{1}{2}$ is because this second choice only occurs half the time.

Theorem 2. *The only H satisfying the three above assumptions is of the form:*

$$H = -K \sum_{i=1}^{n} p_i \log p_i$$

where K is a positive constant.

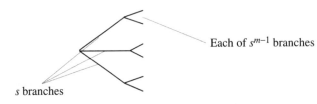

Each of s^{m-1} branches

s branches

Figure 14.3. The tree diagram for equally likely choices.

We need not concern ourselves about the meaning of the most famous legacy of Chebychev's student, Andrei Markov (1856–1922). Passing the reference by, the first two conditions are intuitively reasonable, but the third demands more care. Firstly, to develop a feeling for what is happening, suppose there is only one event possible, then there is no uncertainty and we would want $H(p_1) = H(1)$ to be 0. Now suppose that there are two possibilities, then we can write $H(p_1, p_2) = H(p, 1-p)$. If p is close to 0 or 1, there is little uncertainty and we intuitively feel that in these cases H should be near 0, with the maximum uncertainty achieved when $p = \frac{1}{2}$. We would then reasonably expect a graph of $f(p) = H(p, 1-p)$ to look something like that in Figure 14.2. The third condition carries with it the usual meaning of tree diagrams and is best looked at in two stages. If all n choices are equally likely, write (as Shannon did)

$$A(n) = H\left(\frac{1}{n}, \frac{1}{n}, \frac{1}{n}, \ldots, \frac{1}{n}\right).$$

Now suppose that $n = s^m$ for some positive integers s and m, then the choice can be made in two stages, as in Figure 14.3.

Which makes

$$A(s^m) = A(s) + \frac{1}{s}A(s^{m-1}) \times s = A(s) + A(s^{m-1}),$$

using the fact that we can make the remaining s^{m-1} choices in s equally likely ways, each with a probability of $1/s$. Repeating the process results in $A(s^m) = mA(s) + A(1)$ and since $A(1) = H(1) = 0$, we have that $A(s^m) = mA(s)$ and we can begin to discern properties of logarithms appearing.

We will continue to follow Shannon's reasoning as he develops the full logarithmic behaviour of this equally likely form of uncertainty and from that establishes the result for its most general form.

For an arbitrary large chosen positive integer n, choose a positive integer m and positive integers s and t so that $s^m \leqslant t^n \leqslant s^{m+1}$, which makes (to any base)

$$\log s^m \leqslant \log t^n \leqslant \log s^{m+1},$$

$$m \log s \leqslant n \log t \leqslant (m+1) \log s,$$

$$\frac{m \log s}{n \log s} \leqslant \frac{n \log t}{n \log s} \leqslant \frac{(m+1) \log s}{n \log s},$$

$$\frac{m}{n} \leqslant \frac{\log t}{\log s} \leqslant \frac{m}{n} + \frac{1}{n},$$

$$0 \leqslant \frac{\log t}{\log s} - \frac{m}{n} \leqslant \frac{1}{n},$$

which we will write as

$$\left| \frac{\log t}{\log s} - \frac{m}{n} \right| \leqslant \frac{1}{n}.$$

He then establishes a similar inequality for the function A.

We have that $A(s^m) \leqslant A(t^n) \leqslant A(s^{m+1})$ and since $A(s^m) = m A(s)$ and $A(t^m) = m A(t)$, we have

$$m A(s) \leqslant n A(t) \leqslant (m+1) A(s),$$

$$\frac{m A(s)}{n A(s)} \leqslant \frac{n A(t)}{n A(s)} \leqslant \frac{(m+1) A(s)}{n A(s)},$$

$$\frac{m}{n} \leqslant \frac{A(t)}{A(s)} \leqslant \frac{m}{n} + \frac{1}{n},$$

and, as before,

$$\left| \frac{A(t)}{A(s)} - \frac{m}{n} \right| \leqslant \frac{1}{n}.$$

Combining these gives

$$\left| \frac{A(t)}{A(s)} - \frac{\log t}{\log s} \right| \leqslant \frac{2}{n}$$

and since n can be taken to be arbitrarily large, this makes

$$\frac{A(t)}{A(s)} = \frac{\log t}{\log s} \quad \text{and} \quad \frac{A(t)}{\log t} = \frac{A(s)}{\log s}$$

for all such s and t, which must mean that $A(t) = K_1 \log t$, for some constant K_1.

The move to the general expression for H is made in the following way.

Suppose that we have n different choices c_r, with each choice occurring n_r times $1 \leqslant r \leqslant n$. This means that

$$n = \sum_{r=1}^{n} n_r \quad \text{and} \quad p_r = \frac{n_r}{n}.$$

We can think of the available choice in two ways.

The possibilities can be listed as shown in Figure 14.4 and the choice be made by considering them to be n possibilities, all equally likely, to give $A(n)$. Or

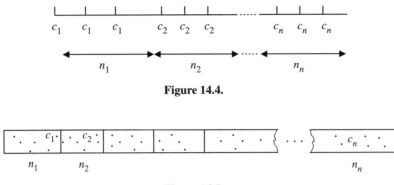

Figure 14.4.

Figure 14.5.

we can group the identical ones together, see which group is chosen and then see which member of that group is chosen, which is represented in Figure 14.5. Now our measure of choice is $H(p_1, p_2, \ldots, p_n) + \sum_{r=1}^{n} p_r A(n_r)$, where the first part relates to the uncertainty of which box is chosen and the second part the uncertainty of which of the equally likely choices are made within a box.

Equating the two forms gives

$$A(n) = H(p_1, p_2, \ldots, p_n) + \sum_{r=1}^{n} p_r A(n_r),$$

which makes

$$K_1 \log n = H(p_1, p_2, \ldots, p_n) + K_1 \sum_{r=1}^{n} p_r \log n_r$$

so

$$H(p_1, p_2, \ldots, p_n) = K_1 \log n - K_1 \sum_{r=1}^{n} p_r \log n_r$$

$$= K_1 \log n \sum_{r=1}^{n} p_r - K_1 \sum_{r=1}^{n} p_r \log n_r$$

$$= K_1 \sum_{r=1}^{n} p_r \log n - K_1 \sum_{r=1}^{n} p_r \log n_r$$

$$= K_1 \sum_{r=1}^{n} p_r \log \frac{n}{n_r} = K_1 \sum_{r=1}^{n} p_r \log p_r.$$

Note that K_1 must be chosen to be negative, since the logarithms are all negative and H must be increasing as a function of n, so write $K = -K_1$, with $K > 0$,

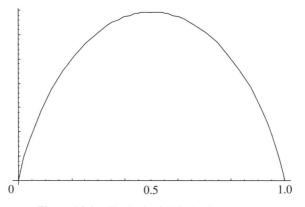

Figure 14.6. The look of H in the 2-state case.

to give

$$H(p_1, p_2, \ldots, p_n) = -K \sum_{r=1}^{n} p_r \log p_r.$$

The choice of a value for K is arbitrary, as is the base of the logarithms, and Shannon's concept of entropy (he points out) '... will be recognized as that of entropy as defined in certain formulations of statistical mechanics...'.

If we perform a small check with the case $n = 2$, choosing $K = 1$ and natural logarithms, we have that $f(p) = H(p, 1 - p) = -p \ln p - (1 - p) \ln(1 - p)$, which looks like the graph in Figure 14.6, and which is what we would have hoped for. Uncertainty is logarithmic—and very important.

14.2 BENFORD'S LAW

Logarithm tables have helped to solve countless problems since Napier's invention of them, and they have created one too, a particularly strange phenomenon that at first sight seems barely plausible, but to which they themselves are the solution. Suppose that an English-speaking student is learning the French language and has a combined English/French and French/English dictionary, split into two halves. It is very likely that the English/French half of the book will be more used than the other and we would expect as time goes by for the book to show uneven signs of wear; there is no surprise here. A book of logarithms is different. If, over time, it is used for a variety of calculations, we would expect its use to be evenly distributed throughout its pages: it isn't.

The distinguished American mathematician and astronomer Simon Newcomb (1835–1909) was made a Foreign Member of The Royal Society on 13 December 1877, exactly the same date as Chebychev was so honoured. We have mentioned Chebychev before and more of his mathematics will be discussed in

Table 14.1. Distribution of first significant digits.

d	Intuitive probability	Suggested probability
1	0.111...	0.301 03
2	0.111...	0.176 09
3	0.111...	0.124 94
4	0.111...	0.096 91
5	0.111...	0.079 18
6	0.111...	0.066 95
7	0.111...	0.057 99
8	0.111...	0.051 15
9	0.111...	0.045 78

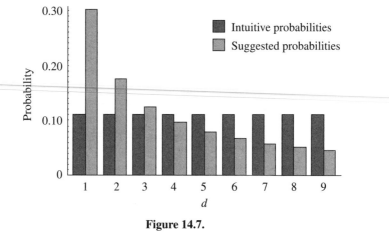

Figure 14.7.

Chapter 15, but Newcomb's offbeat observation has its place here. He noticed that the books of logarithms that he shared with other scientists showed greater signs of use at their beginning than they did at their end. Since log tables are arranged in ascending numeric order, this suggested that more numbers with small rather than large first significant digits were being used for calculation. Yet, all sorts of numbers of all sorts of sizes were being dealt with; why didn't the distribution of their most significant digits even out? Newcomb's investigations led him to the empirical law that the fraction of numbers that start with the digit d is not the intuitively reasonable $\frac{1}{9}$ but the remarkable $\log_{10}(1 + 1/d)$. In 1881 he mentioned the phenomenon in a brief article in the *American Journal of Mathematics* but, without the mathematical justification to support it, it was no more than a curiosity and disappeared from the mathematical landscape—until 1938, when Frank Benford, a physicist at G.E.C. noticed precisely the same thing. The distribution of first significant digits did not appear to be a uniform $\frac{1}{9}$ but that remarkable $\log_{10}(1 + 1/d)$, summarized in Table 14.1 and Figure 14.7, which compare the intuitive and suggested probabilities of the first significant digit appearing.

If the suggested probabilities are the true measure of the frequency of occurrence of naturally occurring numbers, it is small wonder that at some time someone would notice that the front of a book of log tables is about six times more dirty than the back.

To add weight to the hypothesis, he compiled a table of 20 229 numbers, including such wildly disparate categories as the areas of rivers, death rates, baseball statistics, numbers in magazine articles and the street addresses of the first 342 people listed in the book 'American Men of Science'. The table is reproduced in Table 14.2 and is largely consistent with the idea that these seemingly unrelated sets of numbers follow the same first-digit probability pattern as the worn pages of the logarithm tables.

The assertion that distribution of first significant digits is $\log_{10}(1 + 1/d)$ has subsequently become known as Benford's Law. But where is the mathematics to support it? The counterintuitive nature of the law is a phenomenon seen elsewhere in probability theory, perhaps most common is the 'birthday paradox' (which shows us that only 23 people are needed to have the odds of at least two of them having the same birthday in excess of even). Theodore Hill of the Georgia Institute of Technology refers to another when he has his students choose between tossing a fair coin 200 times or faking the results. It is natural for the fakers to mix up the sequence of heads and tails as much as possible but, as he points out, 'the overwhelming odds are that at some point in a series of 200 tosses, either heads or tails will come up six or more times in a row'.

Many sets of numbers certainly do not obey Benford's Law: random numbers at one extreme and numbers that are governed by some other statistical distribution on the other, perhaps Uniform or Normal. It seems that for data to conform to the law they need just the right amount of structure. The last row of averages of the data in the Benford table, with its excellent fit to the law, reveals some of the mystery and it was Hill who saw into it. In 1996 he showed that if distributions are selected at random and random samples are taken from each of these distributions, the significant-digit frequencies of the combined sample would converge to conform to Benford's Law, even though the individual distributions selected may not. Hill calls it the 'random samples from random distributions'. In a sense, Benford's Law is the distribution of distributions!

There are other ways of approaching the phenomenon. If such a law is to be universal, it must for example apply to the base 5 system of counting of the Arawak's of North America, the base 20 system of the Tamanas of the Orinoco and to the Babylonians with their base 60, as well as to the exotic Basque system, which uses base 10 up to 19, base 20 from 20 to 99 and then reverts to base 10.

Table 14.2. Benford's data.

Title	First digit									Samples
	1	2	3	4	5	6	7	8	9	
Rivers, area	31.0	16.4	10.7	11.3	7.2	8.6	5.5	4.2	5.1	335
Population	33.9	20.4	14.2	8.1	7.2	6.2	4.1	3.7	2.2	3259
Physical constants	41.3	14.4	4.8	8.6	10.6	5.8	1.0	2.9	10.6	104
Numbers from newspaper articles	30.0	18.0	12.0	10.0	8.0	6.0	6.0	5.0	5.0	100
Specific heat	24.0	18.4	16.2	14.6	10.6	4.1	3.2	4.8	4.1	1389
Pressure	29.6	18.3	12.8	9.8	8.3	6.4	5.7	4.4	4.7	703
H.P. lost	30.0	18.4	11.9	10.8	8.1	7.0	5.1	5.1	3.6	690
Molecular weight	26.7	25.2	15.4	10.8	6.7	5.1	4.1	2.8	3.2	1800
Drainage	27.1	23.9	13.8	12.6	8.2	5.0	5.0	2.5	1.9	159
Atomic weight	47.2	18.7	5.5	4.4	6.6	4.4	3.3	4.4	5.5	91
$n^{-1}, n^{1/2}$	25.7	20.3	9.7	6.8	6.6	6.8	7.2	8.0	8.9	5000
Design	26.8	14.8	14.3	7.5	8.3	8.4	7.0	7.3	5.6	560
'Readers digest' data	33.4	18.5	12.4	7.5	7.1	6.5	5.5	4.9	4.2	308
Cost data	32.4	18.8	10.1	10.1	9.8	5.5	4.7	5.5	3.1	741
X-ray volts	27.9	17.5	14.4	9.0	8.1	7.4	5.1	5.8	4.8	707
American League	32.7	17.6	12.6	9.8	7.4	6.4	4.9	5.6	3.0	1458
Blackbody	31.0	17.3	14.1	8.7	6.6	7.0	5.2	4.7	5.4	1165
Addresses	28.9	19.2	12.6	8.8	8.5	6.4	5.6	5.0	5.0	342
Mathematical constants	25.3	16.0	12.0	10.0	8.5	8.8	6.8	7.1	5.5	900
Death rate	27.0	18.6	15.7	9.4	6.7	6.5	7.2	4.8	4.1	418
Average	30.6	18.5	12.4	9.4	8.0	6.4	5.1	4.9	4.7	1011
Probable Error (+ve/−ve)	0.8	0.4	0.4	0.3	0.2	0.2	0.2	0.2	0.3	

The law must be base independent. And indeed it is, since base independence of data has been shown to imply Benford's Law.

The units of measurement should not matter either. For example, the fast-disappearing British Imperial system of measurement of length and mass is

12 inches = 1 foot,	16 ounces = 1 pound,
3 feet = 1 yard,	14 pounds = 1 stone,
$5\frac{1}{2}$ yards = 1 pole (or rod, or perch),	2 stones = 1 quarter,
4 poles = 1 chain,	4 quarters = 1 hundredweight,
10 chains = 1 furlong,	20 hundredweights = 1 ton,
8 furlongs = 1 mile.	

(Incidentally, these are nothing more than examples of a finite mixed-base measuring system, as discussed on p. 99. For example, with the length data, suppose that we have the imperial distance of 7 miles, 5 furlongs, 3 chains, 1 pole, 2 yards, 1 foot and 11 inches. In miles this is the expression

$$7 + 5 \times \frac{1}{8} + 3 \times \frac{1}{8} \times \frac{1}{10} + 1 \times \frac{1}{8} \times \frac{1}{10} \times \frac{1}{4} + 2 \times \frac{1}{8} \times \frac{1}{10} \times \frac{1}{4} \times \frac{1}{5\frac{1}{2}}$$

$$+ 1 \times \frac{1}{8} \times \frac{1}{10} \times \frac{1}{4} \times \frac{1}{5\frac{1}{2}} \times \frac{1}{3} + 11 \times \frac{1}{8} \times \frac{1}{10} \times \frac{1}{4} \times \frac{1}{5\frac{1}{2}} \times \frac{1}{3} \times \frac{1}{12}$$

$$= 7 + \frac{1}{8}\left(5 + \frac{1}{10}\left(3 + \frac{1}{4}\left(1 + \frac{1}{5\frac{1}{2}}\left(2 + \frac{1}{3}\left(1 + \frac{1}{12}(11)\right)\right)\right)\right)\right)$$

$$= [7; 5, 3, 1, 2, 1, 11] = 7.6672 \text{ miles.})$$

Euler's manuscript, 'Meditations upon experiments made recently on the firing of a canon', concerned a series of seven experiments carried out in 1727 and which forever cast the letter e for the base of the natural logarithm; in it he measured the cannon ball's diameter in 'scruples of Rhenish feet'. Surely the same cannon balls would or would not conform to Benford's Law whether their diameters be measured by the English Imperial system, Euler, or our modern metric system or indeed by any other system of measurement. The same point can be made for their masses too. In 1961, Roger Pinkham, a mathematician then at Rutgers University in New Brunswick, proved just that: scale invariance did imply Benford's Law. It is this fact that we will focus on and show how such a result can be established.

A change of units is achieved by multiplying by some scaling number and before we immerse ourselves in the mathematics, we can get a feel for the phenomenon by seeing what happens when we do just that in a particular case. Suppose that we take a hypothetical set of 100 'canon balls' of diameters 1–100 scruples of Rhenish feet, order them descendingly by size and plot order against diameter, to arrive at Figure 14.8(a). Now we change units by multiplying each

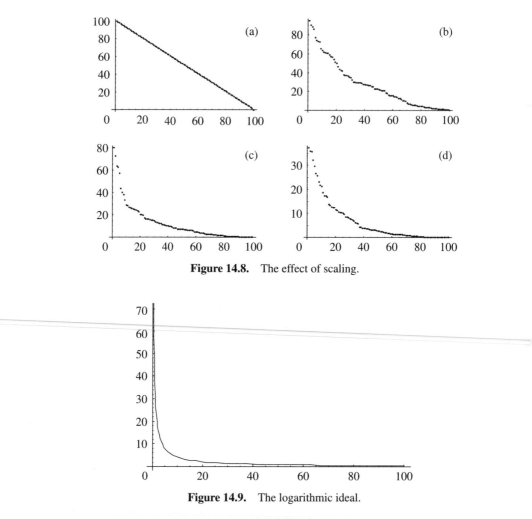

Figure 14.8. The effect of scaling.

Figure 14.9. The logarithmic ideal.

diameter by a random number (in this case, between 0 and 1), and then again, and once more, re-ordering to get Figure 14.8(b)–(d).

The same shapes result from any scalings and the concavity of the resulting curves forces bigger numbers to become more rare. The eye encourages the thought that the plots are approximating some limiting curve. Which curve? Figure 14.9 is a scaled plot of $\log_{10}(1 + 1/\text{diameter})$—which makes one think.

More specifically, consider first significant digits, uniformly distributed, and then suppose that we change the units by multiplying by 2. The first significant digits of the data after the rescaling are given in Table 14.3, which gives rise to the bar chart in Figure 14.10. Equally likely digits are not scale invariant.

Table 14.3.

	Effect of multiplication by 2				
Interval	[1, 1.5)	[1.5, 2)	[2, 2.5)	[2.5, 3)	[3, 3.5)
First significant digit after ×2	2	3	4	5	6
Interval	[3.5, 4)	[4, 4.5)	[4.5, 5)	[5, 10)	
First significant digit after ×2	7	8	9	1	

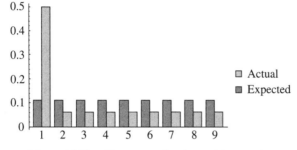

Figure 14.10. The expected and actual frequencies
(distribution of first significant digits).

Now to some mathematics; we will give a statistical definition of scale invariance and use it to show that scale invariance does indeed imply Benford's Law.

We need the ideas of the probability density function $\varphi(x)$ and the cumulative density function $\Phi(x)$ of a continuous random variable. These definitions are the usual

$$P(a \leqslant X \leqslant b) = \int_a^b \varphi(x) \, dx,$$

where $\Phi(x) = P(X \leqslant x) = \int^x \varphi(x) \, dx$ and therefore $d\Phi(x)/dx = \varphi(x)$.

We will say that a random variable X is scale invariant if the probabilities that it lies in any interval before and after scaling (i.e. multiplying) by any factor (say $1/a$) are the same, not worrying about the details of any domains of definition. If we fix on a lower limit and allow the upper limit to vary, we could write this as

$$P(\alpha < X < x) = P\left(\alpha < \frac{1}{a}X < x\right) = P(a\alpha < X < ax),$$

which means that

$$\Phi(ax) - \Phi(a\alpha) = \Phi(x) - \Phi(\alpha) \quad \text{or} \quad \Phi(ax) = \Phi(x) + K_a \quad \text{for all } a.$$

If we differentiate both sides of the above identity with respect to x, we get $a\varphi(ax) = \varphi(x)$ and therefore $\varphi(ax) = (1/a)\varphi(x)$.

151

Now let Y be the random variable $Y = \log_b X$ with $\psi(y)$ and $\Psi(y)$ defined analogously. Then

$$\Psi(y) = P(Y \leqslant y) = P(\log_b X \leqslant y) = P(X \leqslant b^y) = \Phi(b^y) = \Phi(x).$$

This means that

$$\psi(y) = \frac{d}{dy}\Psi(y) = \frac{d}{dy}\Phi(x)$$

$$= \frac{d}{dx}\Phi(x) \times \frac{dx}{dy}$$

and

$$\psi(y) = \varphi(x) \times \frac{dx}{dy} = x\varphi(x)\ln b,$$

so

$$\psi(\log_b x) = \varphi(x) \times \frac{dx}{dy} = x\varphi(x)\ln b,$$

which means that

$$\psi(\log_b ax) = ax\varphi(ax)\ln b.$$

Using the scale invariance we then have

$$\psi(\log_b ax) = ax\varphi(ax)\ln b$$

$$= ax\frac{1}{a}\varphi(x)\ln b$$

$$= x\varphi(x)\ln b$$

$$= \psi(\log_b x).$$

Therefore,

$$\psi(\log_b x + \log_b a) = \psi(\log_b x)$$

and

$$\psi(y + \log_b a) = \psi(y).$$

Since a can be chosen to be anything we wish, $\psi(y)$ repeats itself over arbitrary intervals and it can only be that it is constant. The logarithm of a scale-invariant variable has a constant probability density function.

We can now relate this to the first-digit phenomena by expressing the numbers in scientific notation $x \times 10^n$, where $1 \leqslant x < 10$, the first significant digit d of the number is simply the first digit of x. As we scale the number, we scale x, adjusting its value modulo 10. In this way, we can always think that $1 \leqslant x < 10$ whether scaled or not and if we take the base of the logarithms to be

Table 14.4. Second digit probabilities.

d	Theoretical probability	Actual probability
0	0.1	0.119 68
1	0.1	0.113 89
2	0.1	0.108 82
3	0.1	0.104 33
4	0.1	0.100 31
5	0.1	0.096 67
6	0.1	0.093 37
7	0.1	0.090 35
8	0.1	0.087 57
9	0.1	0.084 99

10, $y = \log_{10} x$ will have a constant probability density function of 1 defined on $[0, 1]$. Therefore, assuming the scale invariance above and for $n \in \{1, \ldots, 9\}$,

$$
\begin{aligned}
P(d = n) &= P(n \leqslant x < n + 1) \\
&= P(\log_{10} n \leqslant \log_{10} x < \log_{10}(n + 1)) \\
&= P(\log_{10} n \leqslant y < \log_{10}(n + 1)) \\
&= (\log_{10}(n + 1) - \log_{10} n) \times 1 \\
&= \log_{10} \left(\frac{n + 1}{n} \right) = \log_{10} \left(1 + \frac{1}{n} \right),
\end{aligned}
$$

which is Benford's Law.

The analysis can be extended to look at the frequency of subsequent digits in the data. For example, if we write the number as $x_1 x_2 \times 10^n$, where $10 \leqslant x_1 x_2 \leqslant 99$, and define the random variable X accordingly, we get

$$
\begin{aligned}
P(\text{1st significant digit is } x_1 \text{ and the second is } x_2) \qquad \\
= P(x_1 x_2 \leqslant X < x_1 x_2 + 1) \\
= \log_{10} \left(1 + \frac{1}{x_1 x_2} \right).
\end{aligned}
$$

Extending the argument gives

$$
P(\text{second digit is } x_2) = \sum_{r=1}^{9} \log_{10} \left(1 + \frac{1}{x_r x_2} \right),
$$

etc. Table 14.4 shows the full set of probabilities for the appearance of second digits, with 0 now a possible value.

Table 14.5.

First digits of the first 1000 Fibonacci numbers									
Digit	1	2	3	4	5	6	7	8	9
Frequency	301	177	125	96	80	67	56	53	45
Percentage	30	18	13	10	8	7	6	5	5

Using the standard result of conditional probability that

$$P(A \mid B) = P(A \text{ and } B)/P(B)$$

we have

$P(\text{second significant digit is } x_2 \mid \text{first significant digit is } x_1)$

$$= \log_{10}\left(1 + \frac{1}{x_1 x_2}\right) \Big/ \log_{10}\left(1 + \frac{1}{x_1}\right).$$

So, for example, the probability that the second digit of a number is 5 given that its first digit is 6 is

$$\frac{\log_{10}(1 + \frac{1}{65})}{\log_{10}(1 + \frac{1}{6})} = 0.0990,$$

whereas if it started with 9 the probability is

$$\frac{\log_{10}(1 + \frac{1}{95})}{\log_{10}(1 + \frac{1}{9})} = 0.0994.$$

The most likely start to a number turns out to be 10, with a probability of

$$\frac{\log_{10}(1 + \frac{1}{10})}{\log_{10}(1 + \frac{1}{1})} = 0.1375.$$

Having made an appearance, 0 is the most common second digit, but the probabilities are beginning to level out and are nearer the uniform $\frac{1}{10}$ that intuition suggests should be the case; as we move along the number the distribution does approach uniformity and intuition is eventually right.

As we have seen, all manner of diverse data conform to the law. Table 14.5 suggests that the Fibonacci numbers would seem to.

A study by B. Buck and A. C. Merchant of the University of Oxford and S. M. Perez of the University of Cape Town showed that alpha decay half-lives (the time it takes atomic nuclei to lose half their radioactivity by emitting alpha particles) obey Benford's Law both observationally and theoretically. They also remarked that the same behaviour has been observed in monthly electricity bills in the Solomon Islands, the street addresses of eminent American scientists, and the initial digits of 20 of the more important physical constants. Of much more practical interest, financial data seem also to conform; in fact, Benford's Law can be used to test for fraudulent data in income tax returns and other financial reports. Mark Nigrini has made a specialization of this sort of 'forensic auditing', which is called digital analysis. He has written:

> Benford's Law provides auditors with the expected digit frequen-
> cies in tabulated data. By examining the digit and the number fre-
> quencies, auditors can gain data insights that might be missed using
> traditional analytical procedures and sampling methods. The digit
> and number patterns could point to number invention, systematic
> frauds, data errors, or biases in the data. Research is currently
> underway on advanced tests to detect anomalies in data subsets.

One case in which he was involved illustrates his point. Using digital analysis, a company's audit director discovered something odd about the claims being made by the supervisor of the company's healthcare department. The first two digits of the healthcare payments were checked for conformity with Benford's Law, and this revealed a spike in numbers beginning with the digits 65. An audit showed 13 fraudulent cheques for between \$6500 and \$6599 related to fraudulent heart surgery claims processed by the supervisor, with the cheque ending up in her hands. The analysis also uncovered other fraudulent claims worth around \$1 million in total.

This novel and important accounting technique has, of course, heralded Web sites devoted to the production of Benford-compliant data, not for illegal or immoral use, naturally!

14.3 Continued-Fraction Behaviour

A look back at the continued fractions in Chapter 11 might bring to the reader the thought that 1 appears a great deal in the continued fraction form of a number and that, on the whole, the partial quotients are small (although by no means exclusively so, with the 431st of π being 20776 the 5040th of γ 11626 and the mere 5th of π^4 16539); Gauss noticed this too and went much further when he wrote to Laplace on 30 January 1812 about a 'curious problem' that had occupied him for 12 years and which he was unable to resolve to his satisfaction. We will take the reader through what must have been the equivalent of Gauss's reasoning, which led to one of the most remarkable results it is possible to imagine.

Suppose that X is a random variable defined on $\mathbb{R}^+$ and that we write $\{X\}$ for the fractional part of X. If the integer part of X is uniformly distributed, then $P(\{X\} < x) = x, 0 \leqslant x < 1$, but suppose that it is not, then this probability will vary according to the value of X and we would have to divide up the real line, as in Figure 14.11, to get

$$P(\{X\} < x) = \sum_{k=1}^{\infty} (P(X < k + x) - P(X < k)).$$

All well and good, but now we apply this idea to continued fractions. Define ξ_n by

$$\xi_n = a_n + \cfrac{1}{a_{n+1} + \cfrac{1}{a_{n+2} + \cdots}} = a_n + \frac{1}{\xi_{n+1}},$$

in which case, $1/\xi_{n+1}$ is the fractional part of ξ_n. Now let

$$\omega_n(x) = P(\{\xi_n\} < x)$$

$$= \sum_{k=1}^{\infty} (P(\xi_n < k + x) - P(\xi_n < k))$$

$$= \sum_{k=1}^{\infty} \left(P\left(\frac{1}{\xi_n} > \frac{1}{k+x}\right) - P\left(\frac{1}{\xi_n} > \frac{1}{k}\right) \right)$$

$$= \sum_{k=1}^{\infty} \left(\left[1 - P\left(\frac{1}{\xi_n} < \frac{1}{k+x}\right)\right] - \left[1 - P\left(\frac{1}{\xi_n} < \frac{1}{k}\right)\right] \right)$$

$$= \sum_{k=1}^{\infty} \left(P\left(\frac{1}{\xi_n} < \frac{1}{k}\right) - P\left(\frac{1}{\xi_n} < \frac{1}{k+x}\right) \right)$$

$$= \sum_{k=1}^{\infty} \left(P\left(\{\xi_{n-1}\} < \frac{1}{k}\right) - P\left(\{\xi_{n-1}\} < \frac{1}{k+x}\right) \right)$$

$$= \sum_{k=1}^{\infty} \left(\omega_{n-1}\left(\frac{1}{k}\right) - \omega_{n-1}\left(\frac{1}{k+x}\right) \right),$$

and we have a recurrence relation for $\omega_n(x)$; the question is, can we find an explicit formula? An intuitive way forward is to argue that, since the relation holds for all n, if the limit $\omega(x)$ exists as $n \to \infty$, we can reasonably hope for it to satisfy

$$\omega(x) = \sum_{k=1}^{\infty} \left(\omega\left(\frac{1}{k}\right) - \omega\left(\frac{1}{k+x}\right) \right)$$

and, remembering that $\omega(x)$ is the limit of the probability of a fraction being less than x, it should be that $\omega(0) = 0$ and $\omega(1) = 1$, which is where mortals might leave the matter.

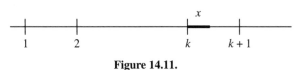

Figure 14.11.

Gauss mentioned in his letter to Laplace that he 'could prove by a very simple argument' that $\omega(x) = \log_2(1+x)$, which brings us to the promised surprising appearance of logarithms. Of course, this does satisfy the two conditions and we will show, as no doubt he did, that it does satisfy the recurrence relation also, but what mysterious thought process he used to arrive at the solution is hard to imagine.

So, if $\omega(x) = \log_2(1+x)$,

$$\sum_{k=1}^{N} \log_2 \left(\frac{k+1}{k} \times \frac{k+x}{k+x+1} \right)$$

$$= \log_2 \prod_{k=1}^{N} \frac{k+1}{k} \times \frac{k+x}{k+x+1}$$

$$= \log_2 \left(\frac{\cancel{2}}{1} \times \frac{1+x}{\cancel{2+x}} \right)\left(\frac{\cancel{3}}{\cancel{2}} \times \frac{\cancel{2+x}}{\cancel{3+x}} \right)\left(\frac{\cancel{4}}{\cancel{3}} \times \frac{\cancel{3+x}}{\cancel{4+x}} \right) \cdots$$

$$\cdots \left(\frac{N+1}{\cancel{N}} \times \frac{\cancel{N+x}}{N+x+1} \right)$$

$$= \log_2 \frac{(1+x)(N+1)}{N+x+1} \xrightarrow[N\to\infty]{} \log_2(1+x).$$

What Gauss could not do was to forge these ideas into the statement

$$P([0; a_1, a_2, a_3, \ldots, a_n] < x) = \omega_n(x) = \log_2(1+x) + \varepsilon_n$$

and therefore rigorously produce what might be thought of as his 'second statistical distribution' (although the first, the ubiquitous 'Gaussian' or 'Normal' or 'Error distribution, which was used by Gauss in 1809 to analyse astronomical data, was used by Laplace in 1783 to investigate errors in measurement and came into being through the work of de Moivre, who in 1733 developed it as an approximation to the Binomial Distribution).

In the end, the problem was solved independently by two mathematicians. In 1928 R. O. Kuzmin showed that, for almost all continued fractions, $\varepsilon_n = O(q^{\sqrt{n}})$, where $0 < q < 1$, and in 1929 Paul Lévy (1886–1971) showed in a completely different way that $\varepsilon_n = O(q^n)$, where $q = 0.7$ and we have error terms that are not only relatively small but asymptotically zero.

157

Table 14.6. Partial quotient distribution for almost all continued fractions.

	For large n, % probabilities for partial quotients								
k	1	2	3	4	5	6	7	8	9+
$P(a_n = k)$	41	17	9	6	4	3	2	2	16

From this incredible result we can find the probability density function of the partial quotients

$$P(a_n = k) = P(k < \xi_n < k+1) = P(\xi_n < k+1) - P(\xi_n < k)$$

$$= \omega_n\left(\frac{1}{k}\right) - \omega_n\left(\frac{1}{k+1}\right)$$

$$\xrightarrow[n \to \infty]{} \log_2\left(1 + \frac{1}{k}\right) - \log_2\left(1 + \frac{1}{k+1}\right)$$

$$= \log_2\left(\frac{(k+1)^2}{k(k+2)}\right) = \log_2\left(\frac{k(k+2)+1}{k(k+2)}\right)$$

$$= \log_2\left(1 + \frac{1}{k(k+2)}\right)$$

which gives rise to Table 14.6. We can check that it is indeed a probability density function:

$$\sum_{k=1}^{N} \log_2\left(1 + \frac{1}{k(k+2)}\right)$$

$$= \sum_{k=1}^{N} \log_2\left(\frac{(k+1)^2}{k(k+2)}\right)$$

$$= \sum_{k=1}^{N} \{2 \log_2(k+1) - \log_2 k - \log_2(k+2)\}$$

$$= \sum_{k=1}^{N} \{\log_2(k+1) - \log_2 k\} + \sum_{k=1}^{N} \{\log_2(k+1) - \log_2(k+2)\}$$

$$= \log_2(N+1) + \log_2 2 - \log_2(N+2)$$

$$= \log_2 2 + \ln\left(\frac{N+1}{N+2}\right) \xrightarrow[N \to \infty]{} \log_2 2 = 1,$$

with the terms of the two series cancelling.

For example, this tells us that in the approximation for γ

$$P(a_n = 11\,626) = \log_2\left(1 + \frac{1}{11\,626 \times 11\,628}\right) \approx 10^{-8}.$$

Table 14.7.

	Frequency of digits in 1000 partial quotients of γ								
k	1	2	3	4	5	6	7	8	9+
a_n	417	168	75	57	41	33	22	19	168
Actual (%)	42	17	8	6	4	3	2	2	17

Table 14.7 provides ample evidence that γ behaves as 'almost any' number, yet e must be exceptional since 1 is the only odd number appearing in the continued-fraction expansion and every even number appears once and only once; evidently, the Golden Ratio φ is exceptional too.

Now that we have a probability distribution, it is natural to ask what is the average of the a_n—and here is another surprise: there isn't one, as we can see from the following argument. By definition, the average value is

$$\sum_{k=1}^{\infty} k P(a_n = k) \xrightarrow[n \to \infty]{} \sum_{k=1}^{\infty} k \log_2 \left(1 + \frac{1}{k(k+2)}\right),$$

which seems fine, but as k becomes large, $k(k+2) \approx k^2$ and

$$\log_2 \left(1 + \frac{1}{k(k+2)}\right) \approx \log_2 \left(1 + \frac{1}{k^2}\right) = \frac{1}{\ln 2} \ln \left(1 + \frac{1}{k^2}\right) \approx \frac{1}{\ln 2} \frac{1}{k^2},$$

which makes

$$\sum_{k,n \text{ large}}^{\infty} k P(a_n = k) \approx \frac{1}{\ln 2} \sum_{k,n \text{ large}}^{\infty} k \times \frac{1}{k^2} = \frac{1}{\ln 2} \sum_{k,n \text{ large}}^{\infty} \frac{1}{k}$$

and the divergent harmonic series makes another surprising (and unwelcome) appearance. Of course, this analysis does not work for φ and e, although it is obvious that the average convergent for φ is 1. It is undefined for e, as we can see if we reason that adding the convergents means adding pairs of 1s, which is linear in n, and the arithmetic series $2 + 4 + 6 + \cdots$, which is quadratic in n; division by n will leave something of the order of n and be divergent.

Even though the arithmetic mean is not properly defined for the a_n, Aleksandr Khinchin (whom we mentioned earlier on p. 140) proved that the geometric mean does converge, and that for almost all numbers $(a_1 a_2 a_3 \cdots a_n)^{1/n} \to \kappa = 2.685\,45\ldots$, which is appropriately known as Khinchin's constant; the plots in Figure 14.12 suggest that γ, π and κ itself obey Khinchin's law.

The geometric mean for φ is obviously 1 and for e it is undefined, which can be seen using Stirling's approximation, which we developed in Chapter 10. Recall that to a first approximation it states that $n! \approx \sqrt{2\pi n} n^n e^{-n}$.

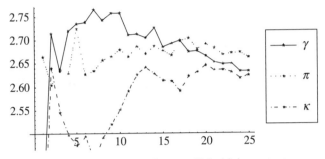

Figure 14.12. The tendency to Khinchin's constant.

An examination of the pattern in the continued fraction form of e shows that

$$\prod_{k=1}^{3n-1} a_k = \prod_{k=1}^{3n} a_k = \prod_{k=1}^{3n+1} a_k = 2^n n!,$$

so if $N = 3n$,

$$\left(\prod_{k=1}^{N} a_k\right)^{1/N} = (2^{N/3}(\tfrac{1}{3}N)!)^{1/N} \approx \left(2^{N/3}\sqrt{2\pi\tfrac{1}{3}N}(\tfrac{1}{3}N)^{N/3}e^{-N/3}\right)^{1/N}$$

$$= (\sqrt{2\pi})^{1/N}(\tfrac{1}{3}N)^{1/(2N)}(\tfrac{1}{3}N)^{1/3}2^{1/3}e^{-1/3}$$

$$\xrightarrow[N\to\infty]{} 1 \times 1 \times \left(\frac{2}{3e}\right)^{1/3} N^{1/3} = 0.6259\cdots N^{1/3},$$

which diverges to ∞.

The Khinchin result can be pushed a little further if we recall the use of the harmonic series in measuring the independence of record events, as discussed on p. 125. With almost all continued fractions the geometric means of the a_n will fluctuate around and home in on κ and it makes sense to record the n for which the geometric mean of the a_n are the 'best yet' in approximating κ; for example, with κ itself the sequence starts

$$1, 2, 3, 15, 23, 26, 81, 104, 109, 111, 120, 127, 135, 136, 141, 142,$$
$$144, 145, 146, 147, 148, 5920, 5943, 8381, 8401, 89\,953, 91\,368, \ldots.$$

So, over $91\,368$ convergents we have 27 records and $H_{91\,368} = 12$; the same calculations for π show that there are 27 records up to $4\,497\,058$ convergents and $H_{4\,497\,058} = 16$, which suggests an unsurprising dependence among the convergents in both cases.

If we recall a definition of the statistical independence of two events A and B is $P(A \text{ and } B) = P(A) \times P(B)$, we can quantify this suspicion since, using

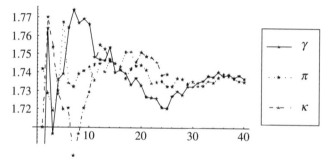

Figure 14.13. Another Khinchin constant.

the distribution above, it can be shown that the partial quotients are 'weakly dependent' in that

$$P(a_n = r \text{ and } a_{n+k} = s) = P(a_n = r) \times P(a_{n+k} = s) \times (1 + O(q^k)),$$

where $0 < q < 1$.

The curious $2.685\,45\ldots$ that is Khinchin's constant is in fact

$$\prod_{r=1}^{\infty} \left(1 + \frac{1}{r(r+2)}\right)^{\ln r / \ln 2},$$

which Khinchin identified by proving the general result that, if $f(r)$ is a sufficiently well-behaved function defined on positive integers, then

$$\frac{1}{n} \sum_{r=1}^{n} f(a_r) \xrightarrow[n \to \infty]{} \frac{1}{\ln 2} \sum_{r=1}^{\infty} f(r) \ln \left(1 + \frac{1}{r(r+2)}\right).$$

His constant results from taking $f(r) = \ln r$. Of course, all manner of choices of $f(r)$ are available and picking $f(r) = 1/r$, generalizing the harmonic mean from p. 121 and rewriting gives

$$H_n = \frac{n}{\sum_{r=1}^{n}(1/a_r)} \xrightarrow[n \to \infty]{} \frac{\ln 2}{\sum_{r=1}^{\infty}(1/r)\ln(1 + 1/r(r+2))}$$

$$= 1.745\,405\,68\ldots$$

and we have the harmonic mean of almost all continued fractions also converging to a limit independent of the fraction itself, as we can see in Figure 14.13. In this case the limit appears to have no name attached to it; perhaps we should call it Khinchin's second constant.

Problems with Primes

Mathematicians have tried in vain to this day to discover some order in the sequence of prime numbers, and we have reason to believe that it is a mystery into which the mind will never penetrate.

Leonard Euler

15.1 SOME HARD QUESTIONS ABOUT PRIMES

Prime numbers have appeared several times in this book. Their study has long held centre stage in number theory and their behaviour, at times seemingly so undisciplined, can sometimes appear determined by an unknown, powerful authority unwilling to disclose its design. The leading quotation makes evident the great Euler's frustration; Erdös, paraphrasing Einstein, said 'God may not play dice with the Universe, but there's something strange going on with the prime numbers!' and R. C. Vaughan spoke for many when he said, 'It is evident that the primes are randomly distributed but, unfortunately, we don't know what random means.' Three among so very many quotations made across the centuries which together encapsulate the wonder in which the behaviour of primes is held.

Of all the questions that can be asked, perhaps the three most fundamental are the following.

(1) Is a given number prime?

(2) How many primes are there less than or equal to a given number x?

(3) What is the xth prime, p_x?

They are easily answered for small numbers: 101 is prime, the 50th prime is 229 and there are 1229 primes less than 10 000 but the going gets much tougher as the numbers get bigger and, after all, we know that there is an infinity of primes. Is 252 097 800 623 prime? How many primes are there less than 100 000 000 000 000 000 000? What is the 1 000 000 000 000 000 000th prime?

These questions are not nearly so straightforward to answer—and these are still 'small' numbers.

We will not dwell on the first question, but the reader will need little convincing that the methods to test for primeness of large numbers are far more subtle than trying to divide the candidate by all primes less than its square root. The question is linked to finding the largest known prime, a search that is inevitably focused on Mersenne primes, mentioned on p. 116, named after the 16th century monk Marin Mersenne and which are of the form $2^p - 1$ with p prime, since for such candidates something called the Lucas–Lehmer test is available. On 5 December 2001 the Great Internet Mersenne Prime Search (GIMPS) initiative found the latest such monster: $2^{13\,466\,917} - 1$ is prime. The number has 4 053 946 digits!

To approach the second question—and, it will turn out, the third too—we will adopt the standard notation $\pi(x)$ for the function which gives the number of primes less than or equal to x, which is known as the 'prime counting function'; so, remembering that 2 is prime and 1 is not, $\pi(3) = 2$, $\pi(17) = 7$ and $\pi(22) = 8$, etc. Clearly, $\pi(x)$ is an increasing step-function of x and since there is an infinite number of primes, we know that $\pi(x) \to \infty$ as $x \to \infty$, but how quickly? The identification of the precise nature of $\pi(x)$ has become known as the Prime Number Theorem and through it we will see how intimately the primes are linked to logarithms and how very remarkable that fact is. In the words of L. J. Goldstein,

> The history of the Prime Number Theorem provides a beautiful example of the way in which great ideas develop and interrelate, feeding upon one another ultimately to yield a coherent theory which rather completely explains observed phenomena.

15.2 A Modest Start

A closer look at Euclid's argument proving the infinity of primes allows us a first (and very poor) lower bound on the size of $\pi(x)$. Although we used the first n primes in the original argument on p. 28, it is clear that $P_n = 1 + p_1 p_2 \cdots p_n$ can be constructed from any set of n primes and of course may or may not itself be prime; whatever the case, let p_{n+1} be the smallest prime dividing P_n, then $p_{n+1} \leqslant P_n = 1 + p_1 p_2 \cdots p_n \leqslant 2p_1 p_2 \cdots p_n$, a huge and costly overestimate. Now suppose that we take $p_1 = 2$, then $p_2 \leqslant 2p_1 = 2 \times 2 = 2^2$, $p_3 \leqslant 2p_1 p_2 = 2 \times 2 \times 2^2 = 2^4$, $p_4 \leqslant 2p_1 p_2 p_3 = 2 \times 2 \times 2^2 \times 2^4 = 2^8$ and in general $p_{n+1} \leqslant 2^{2^n}$, which is an estimate for the size of the nth prime. Since for all $k = 1, 2, \ldots, n$, $p_k < p_{n+1}$, it must be that $p_1, p_2, p_3, \ldots, p_n, p_{n+1} \leqslant 2^{2^n}$. This means that $\pi(2^{2^n}) \geqslant n + 1$. Now write $x = 2^{2^n}$ and so $n = \log_2 \log_2 x$ to get $\pi(x) \geqslant \log_2 \log_2 x + 1 > \log_2 \log_2 x$. Clearly, this inequality will also hold for all $x \geqslant 2^{2^n}$ and we have the bound $\pi(x) > \log_2 \log_2 x$ and a first, early appearance of logarithms.

We can improve matters with a bit more work.

Factorials and the Floor function can be used to count the contribution to $n!$ of each of its prime factors, which in turn has deeper implications, as we will see later. To get an idea of what is happening, consider, for example, $10! = 3\,628\,800 = 2^8 \times 3^4 \times 5^2 \times 7$; 2 appears 8 times, 3 appears 4 times, etc., and, of course, in theory we can factor specific higher factorials and answer the same question, but it is neater and far more practical to consider the general case. In the preliminaries to the co-prime proof in Chapter 8 we noted that there are $x = \lfloor N/r \rfloor$ numbers up to and including N which have r as a divisor. So, for a given prime $p < n$, there are $\lfloor n/p \rfloor$ integers up to and including n, which are divisible by p and therefore p appears in $n!$ precisely $\lfloor n/p \rfloor$ times. Similarly, p^2 appears in $n!$ precisely $\lfloor n/p^2 \rfloor$ times, p^3 appears $\lfloor n/p^3 \rfloor$ times and so on to p^k appears $\lfloor n/p^k \rfloor$ times, where $p^{k+1} > n$. The total exponent of p in $n!$ can then be conveniently expressed as

$$e_p(n!) = \sum_{r=1}^{\infty} \left\lfloor \frac{n}{p^r} \right\rfloor,$$

where the terms of the seemingly infinite series are zero for $r \geqslant k+1$.

This means that

$$n! = \prod_{p \leqslant n} p^{e_p(n!)} = \prod_{p \leqslant n} p^{\sum_{r=1}^{\infty} \lfloor n/p^r \rfloor},$$

a result attributed to Legendre, whom we saw contribute to the theory of the Gamma function and whom we will meet again later in the chapter.

It is this expression that we will use to estimate $\pi(x)$, but before we do we will take a quick look at its contribution to the solution of a well-known problem, since there is no added cost in doing so: how many zeros end a given factorial? For example, we see from above that 10! ends with just two zeros. To answer this in a systematic way we can use the above result to establish how many times 2 and 5 each appear in 10! and then take the smaller of the two numbers to give the number of ways that $10 = 2 \times 5$ appears and therefore in how many zeros the number ends.

We have then that 2 appears

$$\left\lfloor \frac{10}{2} \right\rfloor + \left\lfloor \frac{10}{2^2} \right\rfloor + \left\lfloor \frac{10}{2^3} \right\rfloor = 5 + 2 + 1 = 8$$

times and 5 appears

$$\left\lfloor \frac{10}{5} \right\rfloor = 2$$

165

Table 15.1. A comparison of the estimates.

x	$\pi(x)$	$\log_2 \log_2 x$	$\dfrac{1}{n} \log_2 n!$
10^6	78 498	4.32	18.49
10^7	664 579	4.54	21.8
10^8	5 761 455	4.73	25.1
10^9	50 847 534	4.90	28.5
10^{10}	455 052 511	5.05	31.8
10^{11}	4 118 054 813	5.20	35.1
10^{12}	37 607 912 018	5.32	38.4
10^{13}	346 065 536 839	5.43	41.7

times; 10 therefore appears 2 times and 10! must end with two zeros, as we can see from the direct calculation above. Put to greater use, for 1000!, 2 appears

$$\left\lfloor \frac{1000}{2} \right\rfloor + \left\lfloor \frac{1000}{2^2} \right\rfloor + \left\lfloor \frac{1000}{2^3} \right\rfloor + \cdots + \left\lfloor \frac{1000}{2^9} \right\rfloor$$
$$= 500 + 250 + 125 + 62 + 31 + 15 + 7 + 3 + 1 = 994$$

times and 5 appears

$$\left\lfloor \frac{1000}{5} \right\rfloor + \left\lfloor \frac{1000}{5^2} \right\rfloor + \left\lfloor \frac{1000}{5^3} \right\rfloor + \left\lfloor \frac{1000}{5^4} \right\rfloor = 200 + 40 + 8 + 1 = 249$$

times and so 1000! end with 249 zeros. It is, of course, the number of times that 5 appears that determines the number of zeros.

To apply Legendre's result to estimate $\pi(x)$ we do the following,

$$e_p(n!) = \left\lfloor \frac{n}{p} \right\rfloor + \left\lfloor \frac{n}{p^2} \right\rfloor + \left\lfloor \frac{n}{p^3} \right\rfloor + \cdots,$$

where the series eventually terminates. We can find an upper bound for $e_p(n!)$ by removing the $\lfloor\ \rfloor$ function and allowing the resulting geometric series to extend to infinity to get

$$e_p(n!) < \frac{n}{p} + \frac{n}{p^2} + \frac{n}{p^3} + \cdots = \frac{n}{p}\left(1 + \frac{1}{p} + \frac{1}{p^2} + \cdots\right) = \frac{n}{p}\frac{1}{1 - 1/p} = \frac{n}{p - 1},$$

which makes $p^{e_p(n!)} < p^{n/(p-1)}$. Since for any number $n \geqslant 2$, $n \leqslant 2^{n-1}$, we have that $p^{e_p(n!)} < p^{n/(p-1)} < (2^{p-1})^{n/(p-1)} = 2^n$ and $n! < (2^n)^{\pi(n)} = 2^{n\pi(n)}$. Taking logs to the base 2 we have that $n\pi(n) > \log_2 n!$ and $\pi(n) > (1/n)\log_2 n!$, our new estimate. This takes a bit of calculating for large n, but since we have Stirling's approximation we can estimate well enough for our

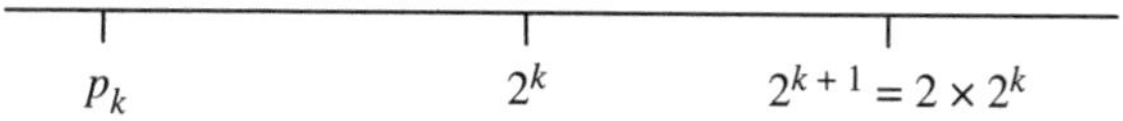

| p_k | 2^k | $2^{k+1} = 2 \times 2^k$ |

Figure 15.1.

purposes by using only the first term of the approximation and taking base 2 logarithms of each side to get

$$\log_2 n! \approx n \log_2 n - n \log_2 e + \tfrac{1}{2} \log_2 2\pi n = n \log_2 \frac{n}{e} + \tfrac{1}{2} \log_2 2\pi n \approx n \log_2 \frac{n}{e}$$

and we have the estimate

$$\pi(n) > \frac{1}{n} \log_2 n! \approx \log_2 \frac{n}{e}$$

for large n.

We can now compile Table 15.1 to see just how bad these estimates really are. On the bright side, at least they are valid bounds and through these ideas we have exercised some small control over the distribution of primes.

We now have two lower bounds on $\pi(x)$. The argument on p. 164 has already provided an upper bound of the size of the nth prime, and this can be significantly sharpened using the Bertrand Conjecture once more (mentioned on p. 25), since if we write the primes in ascending order as $p_1, p_2, \ldots, p_n$ the conjecture implies that $p_n < 2^n$ (of course, $p_1 = 2 = 2^1$, but the inequality is strict from then on). The easiest way to see this is to use induction, referring to Figure 15.1: suppose that for some k, $p_k < 2^k$, then p_{k+1} lies either in the interval $(p_k, 2^k)$, in which case $p_{k+1} < 2^k < 2^{k+1}$, or it lies to the right of 2^k, in which case it must be the first prime that is guaranteed to be in the interval $(2^k, 2^{k+1})$ and again it must be that $p_{k+1} < 2^{k+1}$ and the induction is complete.

15.3 A SORT OF ANSWER

Of course, what we would like is to find an explicit expression for $\pi(x)$ in terms of x and if we are not too choosy, this is readily accomplished. In fact, there are any number of such formulae and a large class of them relies on a result of number theory known as Wilson's Theorem. In 1770, the Cambridge mathematician Edward Waring (1741–1793) published the work *Meditationes Algebraicae*, in which he announced a number of new results of number theory; foremost among them was the statement that if p is prime, then p divides $(p - 1)! + 1$. He attributed it to his former student (and Senior Wrangler), John Wilson (1741–1793), who posited the result on the basis of empirical evidence. No proof was provided. In the publication, Waring admitted to failure in supplying the proof, adding in the text, 'Theorems of this kind will be very hard to prove, because of the absence of a notation to express prime numbers', a comment which failed to impress the great Gauss, who, on reading it, is said to have uttered 'notationes versus notiones', implying that it was the notion that really mattered, not the notation. In fact, it took only until 1773 for Lagrange to provide the proof of statement (and of its inverse), yet it has passed into mathematical lore as Wilson's Theorem; another example of mathematical serendipity. It is even possible that it should carry the name of the mathematical giant Leibnitz, as in his unpublished posthumous papers there are calculations closely related to the idea.

Assuming the truth of Wilson's Theorem, we can give some sort of answers to the last two questions and do so by referring to an article by C. P. Willan in the December 1964 issue of the Mathematical Association's journal *Mathematical Gazette*, which caused a little flurry of conflicting correspondence over the following three years and for that reason alone deserves our attention.

We have, as a direct consequence of Wilson's Theorem, the function

$$F(n) = \left\{ \cos \pi \left(\frac{(n-1)! + 1}{n} \right) \right\}^2 = \begin{cases} 1, & n = 1 \text{ or } n \text{ prime}, \\ 0, & \text{otherwise}, \end{cases}$$

and, consequently,

$$\pi(x) = -1 + \sum_{n=1}^{x} \left\{ \cos \pi \left(\frac{(n-1)! + 1}{n} \right) \right\}^2.$$

To answer the third question, define the function

$$A_n(a) = \left\lfloor \sqrt[n]{\frac{n}{1+a}} \right\rfloor, \quad n = 1, 2, \ldots, \quad a = 0, 1, 2, \ldots.$$

Since, for $a < n$, $1 \leqslant n/(1+a) \leqslant n$ we have that $1 \leqslant \sqrt[n]{n/(1+a)} \leqslant \sqrt[n]{n} < 2$ and so $1 \leqslant A_n(a) \leqslant 1$, which of course forces $A_n(a) = 1$. Similarly, for $a \geqslant n$, $0 < n/(1+a) < 1$ and so $0 \leqslant A_n(a) \leqslant 0$, which forces $A_n(a) = 0$. In summary, then

$$A_n(a) = \begin{cases} 1, & a < n, \\ 0, & a \geqslant n. \end{cases}$$

We can therefore construct the formula

$$p_x = 1 + \sum_{r=1}^{N} A_x(\pi(r)),$$

where N is any sufficiently large integer. We could conveniently take $N = 2^x$ since $p_x \leqslant 2^x$ for all x. The final formula is a typesetter's nightmare when written in full,

$$p_x = 1 + \sum_{r=1}^{2^x} \left\lfloor \sqrt[x]{x - \sqrt[n]{\sum_{s=1}^{r} \left(\cos \pi \frac{(s-1)! + 1}{s} \right)^2}} \right\rfloor,$$

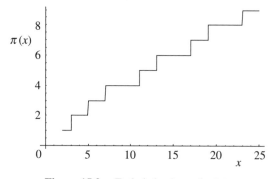

Figure 15.2. Early behaviour of $\pi(x)$.

and it can be quite mysterious to see it at work. For example, Willan gives

$$p_5 = 1 + A_5(\pi(1)) + A_5(\pi(2)) + A_5(\pi(3)) + \cdots + A_5(\pi(32))$$
$$= 1 + A_5(0) + A_5(1) + A_5(2) + \cdots + A_5(11)$$
$$= 1 + 1 + 1 + \cdots + 0 = 11.$$

There are other formulae like them, including another in the same article by Willan not involving $\lfloor\ \rfloor$. The results are novel but it is hard not to feel that this is not really answering the question in the proper spirit, and, anyway, the formulae (and all others derived using the same sort of ideas) are in practice useless for the job for which they are intended.

15.4 PICTURE THE PROBLEM

More realistically, the original question 2 is asking whether an approximation to $\pi(x)$ can be found in the form $\pi(x) = f(x) + \varepsilon_x$ for some easily computable function $f(x)$ and absolute error term ε_x, which we hope not to be too big, and which diminishes asymptotically. To be more precise, we want of the relative error

$$\lim_{x\to\infty} \frac{\pi(x) - f(x)}{\pi(x)} = \lim_{x\to\infty} \frac{\varepsilon_x}{\pi(x)} = 0.$$

So what is this $f(x)$? If we look at the graph of $\pi(x)$ for small x, we see an erratic step function that can do little to boost our confidence in finding it (see Figure 15.2). If we increase the range to $0 \leqslant x \leqslant 100$, the stepped effect is still evident but so is some sort of trend (see Figure 15.3). And for $0 \leqslant x \leqslant 1000$, the trend becomes clearer (see Figure 15.4). Finally, for $0 \leqslant x \leqslant 5000$, we get what appears to the eye near to a straight line; it isn't, of course (see Figure 15.5).

In fact, the curve which the eye superimposes on the graph is concave downwards since, although there is an infinite number of primes, they do become more rare as x increases. The stepped effect is still there, it is simply hidden, and since

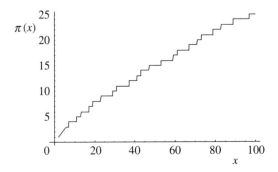

Figure 15.3. A little further on.

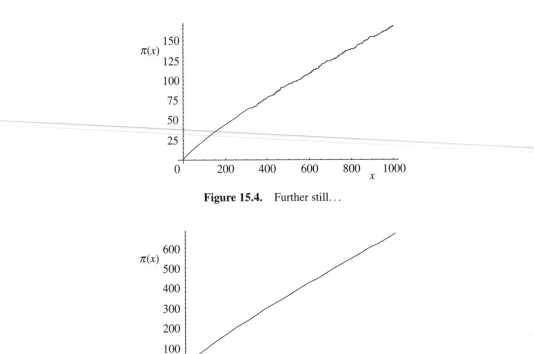

Figure 15.4. Further still...

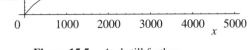

Figure 15.5. And still further...

there are arbitrary distances between primes the 'run' of the steps can be arbitrarily large for a 'rise' of 1. The easiest way to convince oneself of this is to realize that for any positive integer n, the sequence $n! + 2, n! + 3, n! + 4, \ldots, n! + n$ contains no prime.

In 1975, in his inaugural lecture at the university of Bonn, Don Zagier commented:

> There are two facts about the distribution of prime numbers of which I hope to convince you so overwhelmingly that they will be permanently engraved in your hearts. The first is that, despite their simple definition and role as the building blocks of the natural numbers, the prime numbers grow like weeds among the natural numbers, seeming to obey no other law than that of chance, and nobody can predict where the next one will sprout. The second fact is even more astonishing, for it states just the opposite: that the prime numbers exhibit stunning regularity, that there are laws governing their behaviour, and that they obey these laws with almost military precision.

What are these laws that govern the primes' behaviour? In particular, what is that $f(x)$?

15.5 THE SIEVE OF ERATOSTHENES

The Greek scholar, Eratosthenes (276–194 B.C.), was a renowned chronicler of history. He was also chief librarian of the great library of Alexandria and measured the distance along the meridian from there to Assuan, which allowed the size of the Earth to be calculated with remarkable precision. For the mathematician he is remembered more for a device that methodically isolates primes; a device that has become known as his sieve, which allows the creation of a list of primes up to x by knowing the primes up to $\sqrt{x}$, and without a single division.

To use it, we write down all of the integers up to x and then repeatedly cross out every second, third, fifth, etc., integer beyond the first appearance of each for each prime $\leqslant \sqrt{x}$; the remaining uncrossed integers are the primes. Of course, using this new set the whole process can be repeated to find the primes between x and x^2, x^2 and x^4, etc. For example, with $x = 50$ and using the primes 2, 3, 5 and 7 we have Figure 15.6.

Since this isolates the primes, it is small surprise that it can be used to calculate $\pi(x)$ and Daniel Meissel (1826–1895) used it (actually, a refinement of it) to do just that. We mentioned him before on p. 64 and in 1870 he hugely increased the contemporary knowledge by showing that $\pi(10^8) = 5\,761\,455$. In 1893 Bertelsen increased this to $\pi(10^9) = 50\,847\,478$, which was, unfortunately, 56 short of the correct number.

It is interesting to see how the process can be formalized and so realistically begin to deal with large numbers and once again we will use the Floor function and the inclusion–exclusion principle.

Suppose that we fix on an integer and that the list of primes up to $\sqrt{x}$ are $2, 3, 5, \ldots, p_x$. Now modify the process by crossing out the prime as well as

$$
\begin{array}{cccccccccc}
1 & 2 & 3 & \not{4} & 5 & \not{6} & 7 & \not{8} & \not{9} & \not{10} \\
11 & \not{12} & 13 & \not{14} & \not{15} & \not{16} & 17 & \not{18} & 19 & \not{20} \\
\not{21} & \not{22} & 23 & \not{24} & \not{25} & \not{26} & \not{27} & \not{28} & 29 & \not{30} \\
31 & \not{32} & \not{33} & \not{34} & \not{35} & \not{36} & 37 & \not{38} & \not{39} & \not{40} \\
41 & \not{42} & 43 & \not{44} & \not{45} & \not{46} & 47 & \not{48} & 49 & \not{50}
\end{array}
$$

Figure 15.6. Sieve of Eratosthenes.

its multiples. The first sieving by 2 then crosses out $\lfloor \frac{1}{2}x \rfloor$ numbers and we are left with $x - \lfloor \frac{1}{2}x \rfloor$ of them. The second sieving by 3 crosses out all multiples of 3, but it will come across multiples of 6, which have already been eliminated, so we will have remaining $x - \lfloor \frac{1}{2}x \rfloor - \lfloor \frac{1}{3}x \rfloor + \lfloor x/(2 \times 3) \rfloor$.

The reasoning continues for 5, where we have to compensate for multiples of $2 \times 3 \times 5$ having been subtracted once too often; it is really a direct application of inclusion–exclusion. This leaves

$$
x - \lfloor \tfrac{1}{2}x \rfloor - \lfloor \tfrac{1}{3}x \rfloor - \lfloor \tfrac{1}{5}x \rfloor + \left\lfloor \frac{x}{2 \times 3} \right\rfloor + \left\lfloor \frac{x}{2 \times 5} \right\rfloor + \left\lfloor \frac{x}{3 \times 5} \right\rfloor - \left\lfloor \frac{x}{2 \times 3 \times 5} \right\rfloor
$$

numbers.

And so it continues to the prime p_x. We are left with the number 1 and all primes between $\sqrt{x}$ and x, that is, $\pi(x) - \pi(\sqrt{x}) + 1$ numbers and so we have

$$
\pi(x) - \pi(\sqrt{x}) + 1
$$
$$
= x - \lfloor \tfrac{1}{2}x \rfloor - \lfloor \tfrac{1}{3}x \rfloor - \lfloor \tfrac{1}{5}x \rfloor - \cdots
$$
$$
+ \left\lfloor \frac{x}{2 \times 3} \right\rfloor + \left\lfloor \frac{x}{2 \times 5} \right\rfloor + \left\lfloor \frac{x}{3 \times 5} \right\rfloor \cdots - \left\lfloor \frac{x}{2 \times 3 \times 5} \right\rfloor \cdots
$$

with the dots indicating the extension described above.

It is instructive to apply the formula for, say, $x = 100$ to get $\pi(100) = 25$.

15.6 HEURISTICS

We get further by being vaguer. Let's not worry about the Floor function and the duplication and say that about half of the numbers will be divisible by 2 and so we are left with $(1 - \frac{1}{2})x$ after the first round of crossing out. About one-third of those will be divisible by 3, leaving $(1 - \frac{1}{3})(1 - \frac{1}{2})x$. About one-fifth of those will be divisible by 5, leaving $(1 - \frac{1}{5})(1 - \frac{1}{3})(1 - \frac{1}{2})x$, etc. If we repeat this for all of the primes $\leqslant \sqrt{x}$, we will have approximately

$$
\prod_{p \leqslant \sqrt{x}} \left(1 - \frac{1}{p} \right) x
$$

integers remaining, making

$$\pi(x) \approx \prod_{p \leqslant \sqrt{x}} \left(1 - \frac{1}{p}\right) x \qquad \ldots \text{ish.}$$

The error is building and we could do more to keep track of its size, but that would lead us away from the directions in which we wish to travel.

Along one road, recall one of the two Mertens product formulae from p. 109:

$$\lim_{n \to \infty} \frac{1}{\ln n} \prod_{p \leqslant n} \left(1 - \frac{1}{p}\right)^{-1} = e^{\gamma}.$$

We can avoid using the limit notation and reorganize the result to the form

$$\prod_{p \leqslant n} \left(1 - \frac{1}{p}\right) \approx \frac{e^{-\gamma}}{\ln n}$$

for large n. With $n = \sqrt{x}$ this gives the estimate

$$\pi(x) \approx \frac{e^{-\gamma} x}{\ln \sqrt{x}} = 2e^{-\gamma} \frac{x}{\ln x}$$

and the all-important expression $x / \ln x$ has made its first appearance.

Choosing a second (even more bumpy) road, imagine $\pi(x)$ being differentiable for very large x, or approximated accurately by that smooth curve suggested by Figure 15.5, which we will call by the same name, then from above,

$$\pi'(x) \approx \prod_{p \leqslant \sqrt{x}} \left(1 - \frac{1}{p}\right).$$

Now let h be the average interval between primes around $\sqrt{x}$, then, by the definition of tangent, $\pi'(\sqrt{x}) \approx 1/h$. The expression $(\sqrt{x} + h)^2$ is near to x and we will use the approximation

$$\pi'((\sqrt{x} + h)^2) \approx \prod_{p \leqslant \sqrt{x}} \left(1 - \frac{1}{p}\right)\left(1 - \frac{1}{\sqrt{x}}\right) = \left(1 - \frac{1}{\sqrt{x}}\right)\pi'(x),$$

where we are approximating the greatest prime less than $(\sqrt{x} + h)$ by $\sqrt{x}$, which isn't so very terrible for large x.

Now use Taylor's approximation to give

$$\pi'((\sqrt{x} + h)^2) = \pi'(x + 2h\sqrt{x} + h^2) \approx \pi'(x) + 2h\sqrt{x}\pi''(x).$$

Equate these two and simplify to the horrendous differential equation:

$$2x \frac{\pi''(x)}{\pi'(x)} + \pi'(\sqrt{x}) = 0.$$

173

Fortunately, we have a hint already; let us try $\pi(x) = x/\ln x$. The first term becomes

$$\frac{2(2 - \ln x)}{\ln x (\ln x - 1)}$$

and the second

$$-\frac{2(2 - \ln x)}{(\ln x)^2},$$

with the difference between them the difference between $\ln x$ and $(\ln x - 1)$.

The arguments are hardly incapable of criticism but as heuristics they are fine. They have done what is needed of them, which is to point in the right direction for progress. That function $x/\ln x$ does seem to be intimately linked with $\pi(x)$.

15.7 A LETTER

On Christmas Eve 1849, the 72-year-old Gauss wrote a letter to his 'distinguished friend' and former student, the astronomer, Johann Encke (1791–1865). The letter was in response to one from Encke, in which he had shown his own interest in the frequency of the primes and had posited his own estimate for $\pi(x)$, and began,

> Your remarks concerning the frequency of primes were of interest to me in more ways than one. You have reminded me of my own endeavours in this field which began in the very distant past, in 1792 or 1793, after I had acquired the Lambert supplements to the logarithmic tables.

In 1792 Gauss was 15 years old. The fortuitous gift of a table of logarithms and a supplement which contained tables of prime numbers up to 1 million had enabled the young boy to begin the assault on the nature of $\pi(x)$ (compiled by the German–Swiss mathematician Johann Lambert (1728–1777); his name appeared on p. 93 in connection with the theory of continued fractions). Later Gauss would have access to tables of primes up to 3 million. Table 15.2 shows the initial information that the 15-year-old had to work with and on the basis of this very limited evidence it occurred to him that the pattern that was emerging was that for $x = 10^n$,

$$\pi(x) \approx \frac{1}{\alpha \times n} \times x = \frac{1}{\alpha \times \log_{10} x} \times x,$$

where α seems to be a number just over 2—and well he knew that $\ln 10 = 2.30\dots$. The standard laws of logs then produce $\pi(x) \approx x/\ln x$, in keeping with those other heuristic pointers. This gives $f(x) = G(x) = x/\ln x$ and

$$\pi(x) = \frac{x}{\ln x} + \varepsilon_x = G(x) + \varepsilon_x.$$

Table 15.2.

x	$\pi(x)$	Prime density
$10 = 10^1$	4	$1{:}2.5 = 1{:}(2.5 \times 1)$
$100 = 10^2$	25	$1{:}4 = 1{:}(2 \times 2)$
$1\,000 = 10^3$	168	$1{:}5.96 = 1{:}(1.99 \times 3)$
$10\,000 = 10^4$	1\,229	$1{:}8.14 = 1{:}(2.04 \times 4)$
$100\,000 = 10^5$	9\,592	$1{:}10.43 = 1{:}(2.09 \times 5)$
$1\,000\,000 = 10^6$	78\,498	$1{:}12.74 = 1{:}(2.12 \times 6)$

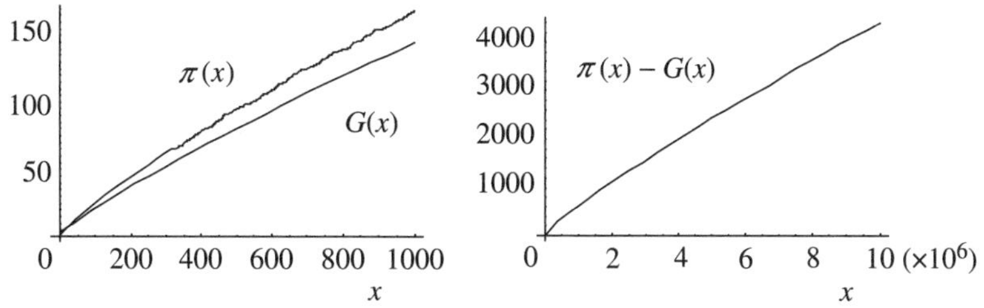

Figure 15.7. Gauss's original estimate.

In Figure 15.7 we have two plots of the early comparison between $\pi(x)$ and $G(x)$. His book of logarithms still survives and has written on its back cover in a young hand 'Primzahlen unter $a(= \infty)a/la$.

In the letter, Gauss referred only to his refined estimate, which came about by localizing the count, considering the number of primes in blocks of 1000 consecutive integers. (There is use of some delightful classical language, with hecatontades for 100, chiliad for 1000 and myriad used in its accurate sense of 10 000.) He wrote that he 'frequently spent an idle quarter of an hour to count another chiliad here and there', which enabled him to average over smaller sub-intervals rather than across the whole interval itself and in the limit 'add up' the primes by integration and so arrive at

$$f(x) = Li(x) = \int_2^x \frac{1}{\ln u}\, du$$

to get the estimate

$$\pi(x) = \int_2^x \frac{1}{\ln u}\, du + \varepsilon_x = Li(x) + \varepsilon_x.$$

And this brings about an appearance of the logarithmic integral function $Li(x)$, which we mentioned on p. 106 and which has become central in the study of the distribution of primes. Predictably, he had failed to publicly announce the

Table 15.3.

x	$\pi(x)$	$Li(x)$	Difference
500 000	41 556	41 606.4	50.4
1 000 000	78 501	78 627.5	126.5
1 500 000	114 112	114 263.1	151.1
2 000 000	148 883	149 054.8	171.8
2 500 000	183 016	183 245.0	229.0
3 000 000	216 745	216 970.6	225.6

Figure 15.8. Gauss's refined estimate.

idea, which was finally published posthumously in 1863 and appears on p. 11, Vol. 10, Part I of his *Werke*, although he did include Table 15.3 in the letter. In every case the prime count is slightly wrong, with the error for the four largest values in favour of the $Li(x)$ estimate.

If we integrate $Li(x)$ by parts twice, we have

$$
Li(x) = \int_2^x \frac{1}{\ln u}\, du = \left[\frac{u}{\ln u}\right]_2^x + \int_2^x \frac{1}{(\ln u)^2}\, du
$$
$$
= \frac{x}{\ln x} + \frac{x}{(\ln x)^2} + \int_2^x \frac{2}{(\ln u)^3}\, du
$$

and a comparison between the two logarithmic estimates, which can be continued as far as we please.

Comparisons for the new estimate of $\pi(x)$ are shown in Figure 15.8. By introducing these estimates, Gauss had established a bridgehead in the battle to harness the behaviour of the primes, but although he worked alone he was not alone in the work. Part way through the letter he commented,

> I was not aware that Legendre had worked on this subject; your letter caused me to look in his *Theorie des Nombres*, and in the second edition I found a few pages on the subject which I must previously have overlooked (or, by now, forgotten).

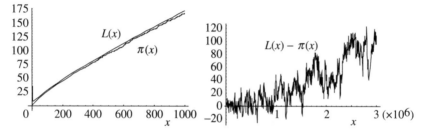

Figure 15.9. Legendre's estimate.

He was referring to Legendre's *Essai sur la Theorie des Nombres*, which originally appeared in 1798 and in an improved second edition in 1808. The original volume contained the proposal that

$$\pi(x) \approx \frac{x}{A \ln x + B}$$

for some constants A and B, which was refined in the second edition, using tables up to 400 000, to the somewhat mysterious

$$f(x) = L(x) = \frac{x}{\ln x - A(x)}$$

and therefore that

$$\pi(x) = \frac{x}{\ln x - A(x)} + \varepsilon_x = L(x) + \varepsilon_x,$$

where $A(x) \approx 1.083\,66$. A formula described by the Norwegian genius Neils Abel (1802–1829), in a letter written in 1823, as the 'most remarkable in the whole of mathematics'. The comparisons are shown in Figure 15.9.

The mysterious $1.083\,66\ldots$ naturally attracted Gauss's interest, as did the fact that up to $3\,000\,000$, $L(x)$ was more accurate than his own $Li(x)$, as we can see from Figure 15.10.

In the letter he recorded the values which $A(x)$ must take for $L(x)$ and $\pi(x)$ to agree over intervals of length $500\,000$ to get values for $A(x)$ of $1.090\,40$, $1.076\,82$, $1.075\,82$, $1.075\,29$, $1.071\,79$, $1.072\,97$. He continued,

> It appears that, with increasing x, the (average) value of $A(x)$ decreases; however, I dare not conjecture whether the limit as x approaches infinity is 1 or a number different from 1. I cannot say that there is any justification for expecting a very simple limiting value.

If we look at the comparisons of $\pi(x)$ with the case $A(x) = 1$, we can see why Legendre would have preferred his strange $1.083\,66$, which must surely have been the result of repeatedly fiddling with the expression. It would be 70 years

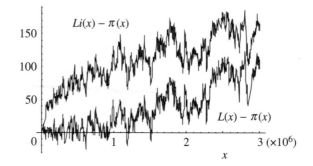

Figure 15.10. The two estimates compared.

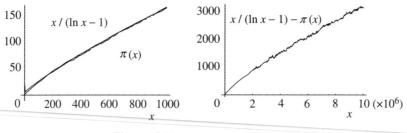

Figure 15.11. The case of $A(x) = 1$.

after Legendre's death before it was proved that, in the long term, Legendre was misled and Gauss was too timid, when it was shown that 1 is in fact the best value.

As to the superiority of $L(x)$ to $Li(x)$, Gauss commented, 'These differences (between $L(x)$ and $\pi(x)$) are even smaller than those from the integral, but they seem to grow faster with x so that it is quite possible they may surpass them'; he was right, eventually they do and it took that same mathematician to prove the fact—but more of that later.

Encke's own estimate is not recorded in the letter but it is interesting to note that Gauss recognized its asymptotic form with,

> By the way, for large x, your formula could be considered to coincide with
> $$\frac{x}{\ln x - (1/2k)},$$
> where k is the modulus of Briggs's logarithms; that is, with Legendre's formula, if we put $A(x) = 1/2k = 1.1513$.

By which he seems to have meant $k = \log_{10} e$.

In summary, we have the tabular comparison in Tables 15.4 and 15.5.

Table 15.4. A table of comparisons.

x	$\pi(x)$	$G(x)$	$L(x)$	$Li(x)$
1 000	168	145	172	178
10 000	1 229	1 086	1 231	1 246
100 000	9 592	8 686	9 588	9 630
1 000 000	78 498	72 382	78 543	78 628
10 000 000	664 579	620 421	665 140	664 918
100 000 000	5 761 455	5 428 681	5 768 004	5 762 209
1 000 000 000	50 847 534	48 254 942	50 917 519	50 849 235
10 000 000 000	455 052 511	434 294 482	455 743 004	455 055 614

Table 15.5. Percentage differences compared with $\pi(x)$.

x	$\%G(x)$	$\%L(x)$	$\%Li(x)$
1 000	−13.8305	2.2027	5.9524
10 000	−11.6569	0.1232	1.3832
100 000	−9.4465	−0.0375	0.3962
1 000 000	−7.7908	0.0576	0.1656
10 000 000	−6.6446	0.0844	0.0510
100 000 000	−5.7759	0.1137	0.0131
1 000 000 000	−5.0988	0.1376	0.0033
10 000 000 000	−4.5617	0.1517	0.0007

15.8 THE HARMONIC APPROXIMATION

One last alternative expression can be extracted from the definition of the harmonic mean of the first x integers. Recall that this has the form

$$H = \frac{x}{\sum_{r=1}^{x} 1/r}$$

and that using the connection between this, ln and γ we have that, for large x, $H \approx x/(\ln x - \gamma)$ and another Legendre-type estimate of $\pi(x)$. This means that the number of primes up to x can be approximated by the harmonic mean of the integers up to x and Figure 15.12 shows this comparison.

The inequality between the harmonic and geometric means established on p. 119 for two numbers can easily be extended to give $H < G$ for any set of numbers. If we consider the set to be the first x integers, this means that

$$\frac{x}{\sum_{r=1}^{x} 1/r} < (1 \times 2 \times 3 \times \cdots \times x)^{1/x} = (x!)^{1/x}$$

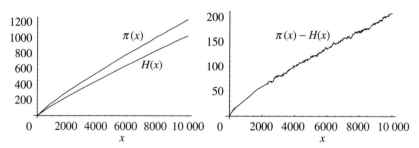

Figure 15.12. The harmonic estimate.

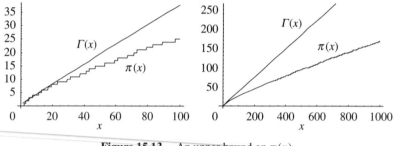

Figure 15.13. An upper bound on $\pi(x)$.

and, once again using the logarithmic approximation to the harmonic series and Stirling's approximation from p. 87 to the factorial, we have

$$\frac{x}{\ln x - \gamma} < (\sqrt{2\pi x} x^x e^{-x})^{1/x} = \frac{(2\pi)^{1/x} x^{1+1/2x}}{e}.$$

Finally, if we allow ourselves the (considerable) luxury of using the Gamma estimate to approximate $\pi(x)$, we have an upper bound on its size, with

$$\pi(x) < \frac{(2\pi)^{1/x} x^{1+1/2x}}{e}$$

for large x.

The graphs in Figure 15.13 show the early and slightly later stages of this (again poor) comparison.

15.9 DIFFERENT—AND YET THE SAME

The expression $\pi(x) = f(x) + \varepsilon_x$, when rewritten as

$$\frac{\pi(x)}{f(x)} = 1 + \frac{\varepsilon_x}{f(x)},$$

allows us to concentrate on the asymptotic comparison of $\pi(x)$ and its approximations and of course we hope for the relative error to diminish to 0 and therefore

$$\lim_{x \to \infty} \frac{\pi(x)}{f(x)} = 1.$$

It is usual to represent such behaviour by the notation $\pi(x) \sim f(x)$.

It is perfectly clear that, if the limit exists,

$$\lim_{x \to \infty} \frac{\pi(x)}{x/\ln x} = \lim_{x \to \infty} \frac{\pi(x)}{x/(\ln x - a)}$$

for any constant a, which makes

$$\pi(x) \sim \frac{x}{\ln x}, \qquad \pi(x) \sim \frac{x}{\ln x - 1.083\,66},$$

$$\pi(x) \sim \frac{x}{\ln x - 1}, \qquad \pi(x) \sim \frac{x}{\ln x - \gamma}$$

equivalent statements in this sense.

That

$$\pi(x) \sim \frac{x}{\ln x} \quad \text{and} \quad \pi(x) \sim \int_2^x \frac{1}{\ln u}\, du$$

are also equivalent takes a bit more work, and we need the help of L'Hôpital's Rule.

One way around, if we assume that

$$\lim_{x \to \infty} \frac{\pi(x)}{x/\ln x} = 1,$$

then

$$\lim_{x \to \infty} \frac{\pi(x)}{\int_2^x (1/\ln u)\, du} = \lim_{x \to \infty} \frac{\pi(x)}{x/\ln x} \cdot \frac{x/\ln x}{\int_2^x (1/\ln u)\, du}$$

$$= 1. \lim_{x \to \infty} \frac{x/\ln x}{\int_2^x (1/\ln u)\, du}$$

and using L'Hôpital's Rule this becomes

$$\lim_{x \to \infty} \frac{(\ln x - x.(1/x))/(\ln x)^2}{1/\ln x} = \lim_{x \to \infty} \left(\frac{\ln x - 1}{(\ln x)^2}. \ln x \right)$$

$$= \lim_{x \to \infty} \frac{\ln x - 1}{\ln x} = 1. \qquad (15.1)$$

The reverse argument is the same. With this established, we can state the celebrated Prime Number Theorem.

> ### Prime Number Theorem
>
> $\pi(x) \sim G(x)$ or equivalently $\pi(x) \sim L(x)$ or $\pi(x) \sim Li(x)$.

We could, of course, add in $\pi(x) \sim x/(\ln x - a)$ for $a = 1$ or otherwise.

15.10 THERE ARE REALLY TWO QUESTIONS, NOT THREE

A little work shows that the Prime Number Theorem is equivalent to estimating the xth prime.

If the Prime Number Theorem is true and if the xth prime is written p_x, then clearly $\pi(p_x) = x$, which intimately associates the growth of $\pi(x)$ with x and p_x with x, and we have

$$\lim_{x \to \infty} \frac{\pi(x)}{x/\ln x} = 1 \Rightarrow \ln \lim_{x \to \infty} \frac{\pi(x)}{x/\ln x} = \ln 1 = 0$$

$$\Rightarrow \lim_{x \to \infty} \ln \frac{\pi(x)}{x/\ln x} = 0$$

$$\Rightarrow \lim_{x \to \infty} (\ln \pi(x) - \ln x + \ln \ln x) = 0$$

$$\Rightarrow \lim_{x \to \infty} \left(\ln x \left(\frac{\ln \pi(x)}{\ln x} + \frac{\ln \ln x}{\ln x} - 1 \right) \right) = 0.$$

Since $\ln x$ is unbounded,

$$\lim_{x \to \infty} \left(\frac{\ln \pi(x)}{\ln x} + \frac{\ln \ln x}{\ln x} - 1 \right) = 0$$

and, since also

$$\lim_{x \to \infty} \frac{\ln \ln x}{\ln x} = 0,$$

we have that

$$\lim_{x \to \infty} \frac{\ln \pi(x)}{\ln x} = 1.$$

So,

$$\lim_{x \to \infty} \frac{\pi(x)}{x/\ln x} \times \lim_{x \to \infty} \frac{\ln \pi(x)}{\ln x} = \lim_{x \to \infty} \frac{\pi(x) \ln \pi(x)}{x} = 1.$$

Now replace x by the xth prime p_x, then, as we have already said, $\pi(p_x) = x$ and the equation becomes

$$\lim_{x \to \infty} \frac{x \ln x}{p_x} = 1$$

and so $p_x \sim x \ln x$.

To show the equivalence we now assume that $p_n \sim n \ln n$ and define n by $p_n \leqslant x < p_{n+1}$. Then $p_n \sim n \ln n$ and $p_{n+1} \sim (n+1) \ln(n+1) \sim n \ln n$ for n large. This means that $x \sim n \ln n$. Also, $\pi(x) = n$, so that $x \sim \pi(x) \ln \pi(x)$. Therefore,

$$
\begin{aligned}
\lim_{x \to \infty} \frac{\pi(x)}{x / \ln x} &= \lim_{x \to \infty} \frac{\pi(x) \ln x}{x} \\
&= \lim_{x \to \infty} \frac{\pi(x) \ln x}{x} \frac{x}{\pi(x) \ln \pi(x)} \\
&= \lim_{x \to \infty} \frac{\ln x}{\ln \pi(x)} = 1.
\end{aligned}
$$

A more delicate argument establishes that $p_x \sim x(\ln x + \ln \ln x - 1)$ and there are improvements to this too. For example, these formulae predict that the one-millionth prime is about $13\,800\,000$ and $15\,400\,000$, respectively; in fact, the one-millionth prime is $15\,485\,863$. In a 1967 paper Rosser and Schoenfeld also showed that

$$
x(\ln x + \ln \ln x - 1.5) < p_x < x(\ln x + \ln \ln x - 0.5)
$$

for $x \geqslant 20$.

15.11 ENTER CHEBYCHEV WITH SOME GOOD IDEAS

So, we have several empirical formulae, essentially identical, but producing different errors in approximating $\pi(x)$—and we have a 'theorem' without a proof. The first major step forward towards achieving a proof was brought about by Chebychev, who used Legendre's result (mentioned on p. 165) and Euler's identity; he also added two functions to his mathematical toolkit.

We can think of the prime counting function being defined by

$$
\pi(x) = \sum_{\substack{p < x \\ p \text{ prime}}} 1,
$$

that is, a step function which increases by 1 whenever a prime is reached. Chebychev generalized this to a weighted prime counting function

$$
\psi(x) = \sum_{\substack{p^r \leqslant x \\ p \text{ prime}}} \ln p,
$$

which increases by $\ln p$ whenever a power of a prime is reached; the sum is interpreted to mean the sum over all primes p such that some positive power of

Table 15.6. Some values of $\psi(x)$.

x	100	200	300	400	500	600	700	800	900	1000
$\psi(x)$	94.04	206.1	299.2	397.8	501.7	593.9	699.0	792.7	897.2	996.7

the prime is less than or equal to x. For example,

$$\psi(20) = (\ln 2 + \ln 3 + \ln 5 + \ln 7 + \ln 11 + \ln 13 + \ln 17 + \ln 19)$$
$$+ (\ln 2 + \ln 3) + (\ln 2) + (\ln 2) = 19.2656\ldots$$

and

$$\psi(30)$$
$$= (\ln 2 + \ln 3 + \ln 5 + \ln 7 + \ln 11 + \ln 13 + \ln 17 + \ln 19 + \ln 23 + \ln 29)$$
$$+ (\ln 2 + \ln 3 + \ln 5) + (\ln 2 + \ln 3) + (\ln 2) = 28.4765\ldots,$$

where the terms are bracketed so that $p < x^{1/r}$ for $r = 1, 2, 3, \ldots$. (A little thought shows that, in fact, $\psi(x) = \ln(\text{g.c.d.}\{1, 2, 3, \ldots, \lfloor x \rfloor\})$.) Chebychev also defined the function $\theta(x) = \sum_{p \leqslant x} \ln p$ and using this and the above bracketing we can easily see that $\psi(x)$ can be written as the finite series ($\theta(y)$ must be zero for $y < 2$)

$$\psi(x) = \theta(x) + \theta(x^{1/2}) + \theta(x^{1/3}) + \theta(x^{1/4}) + \cdots.$$

We can also see that in the two numeric cases detailed above and in Table 15.6, $\psi(x)$ is pretty near to x. Is this a coincidence? Not if the Prime Number Theorem is true, since the statement $\psi(x) \sim x$ is equivalent to it; in fact, we have the

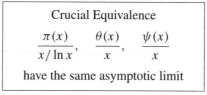

Crucial Equivalence

$$\frac{\pi(x)}{x/\ln x}, \quad \frac{\theta(x)}{x}, \quad \frac{\psi(x)}{x}$$

have the same asymptotic limit

and to prove that Chebychev argued in the following way, which we have taken from A. E. Ingham's treatise, *The Distribution of Prime Numbers* and which we will mention again on p. 188.

Firstly, if $p^r < x$, then r is the maximum value such that $r < \ln x / \ln p$, that is, $r = \lfloor \ln x / \ln p \rfloor$. This means that

$$\psi(x) = \sum_{p \leqslant x} \left\lfloor \frac{\ln x}{\ln p} \right\rfloor \ln p.$$

Now write the three (possibly infinite) limits as L_1, L_2 and L_3, respectively. Then we have the double inequality

$$\theta(x) \leqslant \psi(x) = \sum_{p \leqslant x} \left\lfloor \frac{\ln x}{\ln p} \right\rfloor \ln p \leqslant \sum_{p \leqslant x} \frac{\ln x}{\ln p} \ln p$$

$$= \ln x \sum_{p \leqslant x} 1 = \ln x \pi(x),$$

which means that

$$\frac{\theta(x)}{x} \leqslant \frac{\psi(x)}{x} \leqslant \frac{\pi(x)}{x/\ln x}$$

and this means, taking the limit as $x \to \infty$, $L_2 \leqslant L_3 \leqslant L_1$.

Now suppose that $0 < \alpha < 1$ and that $x > 1$. Then

$$\theta(x) \geqslant \sum_{x^\alpha < p \leqslant x} \ln p$$

$$\geqslant \ln x^\alpha \sum_{x^\alpha < p \leqslant x} 1 = \ln x^\alpha (\pi(x) - \pi(x^\alpha)).$$

Since $\pi(x^\alpha) < x^\alpha$ we have $\theta(x) \geqslant \ln x^\alpha (\pi(x) - x^\alpha)$ and

$$\frac{\theta(x)}{x} \geqslant \frac{\alpha(\pi(x) \ln x - x^\alpha \ln x)}{x}$$

$$= \alpha \left(\frac{\pi(x)}{x/\ln x} - \frac{\ln x}{x^{1-\alpha}} \right).$$

As $x \to \infty$, $\ln x / x^{1-\alpha} \to 0$, which leaves us with $L_2 \geqslant \alpha L_1$ and since this is true for α arbitrarily close to 1, $L_2 \geqslant L_1$. Combine this with the first inequality and we have the result.

By this means, the search for a proof that $\pi(x) \sim x$ can be altered to a search for a proof that $\psi(x) \sim x$. Using such ideas in 1852, in the first of two important papers, Chebychev showed that for arbitrarily large x

$$\int_2^x \frac{du}{\ln u} - \frac{\alpha x}{\ln^n x} < \pi(x) < \int_2^x \frac{du}{\ln u} + \frac{\alpha x}{\ln^n x}$$

for any positive integer n and arbitrarily small $\alpha > 0$, a result which, with $n = 1$, he developed into

$$\frac{\int_2^x (1/\ln u)\, du}{x/\ln x} - \alpha < \frac{\pi(x)}{x/\ln x} < \frac{\int_2^x (1/\ln u)\, du}{x/\ln x} + \alpha$$

and so

$$\lim_{x\to\infty}\frac{\int_2^x(1/\ln u)\,du}{x/\ln x}-\alpha \leqslant \lim_{x\to\infty}\frac{\pi(x)}{x/\ln x}$$

$$\leqslant \lim_{x\to\infty}\frac{\int_2^x(1/\ln u)\,du}{x/\ln x}+\alpha$$

or, using the equivalence (15.1) on p. 181,

$$1-\alpha \leqslant \lim_{x\to\infty}\frac{\pi(x)}{x/\ln x}\leqslant 1+\alpha,$$

which means that if

$$\lim_{x\to\infty}\frac{\pi(x)}{x/\ln x}$$

does exist, then it must be 1. In the same paper he also showed that the relative error in the approximation of $\pi(x)$ by $Li(x)$ is less than 11% for large x but his further attempts to show that it was asymptotically 0 failed.

In his second paper on the subject, dated 1854, he began to close in on the result in that he showed for large x,

$$A_1 < \frac{\pi(x)}{x/\ln x} < A_2,$$

where $0.922\cdots < A_1 < 1$ and $1 < A_2 < 1.105\cdots$. These were major steps forward and they formed a firm base from which to launch attacks on the problem, but the pathway to the ultimate goal seemed irrevocably blocked.

Others tried. None succeeded. Not for another 100 years would a proof be found which is based on 'real' numbers.

A new direction was taken by Dirichlet, whom we have mentioned already on p. 112. In essence, he generalized the definition of the Zeta functions and thereby brought to the mathematical world the L functions, which are a linchpin of modern number theory. We will steer past this elegant and important initiative, but not before mentioning that in 1837 Dirichlet used it to lay to rest the conjecture of Legendre that every arithmetic sequence of integers (with first term co-prime to the common difference) contains an infinite number of primes, and so produce one of the greatest achievements of 19th-century mathematics.

Euler had originally brought analysis into number theory with his identity, Chebychev had Dirichlet had developed the initiative—and then came Riemann with a single idea announced in a single paper.

15.12 ENTER RIEMANN, FOLLOWED BY PROOF(S)

Encke was one of Gauss's distinguished students, Bernhard Riemann (1826–1866) was another. His name has already appeared in these pages but here he plays his most significant role in our story. Shy and introspective, his health never strong, he died of tuberculosis in Italy on the shores of Lake Maggiore; he was 40 years old and mathematically active until the end: the year of his death was also the year in which he was elected to The Royal Society as a Foreign Member. His 'Habilitation' lecture (the final requirement for his acceptance as a lecturer at Göttingen university) had the title 'On the hypotheses that lie at the foundations of geometry', and was delivered on 10 June 1854. It was the third of three titles from which the aged Gauss was to choose and quite the most surprising—and fortunate. Building on Gauss's own ideas, it brought to the mathematical world the clear idea of the intrinsic geometry of space and paved the way for Einstein to formulate his theories of relativity and was to become a classic of mathematics, even though few (other than Gauss) were able to appreciate its profundity at the time. Our interest lies in another paper and the only one that he ever published on number theory. 'On the number of prime numbers less than a given quantity' was submitted to the Berlin Academy of Sciences in 1859 as evidence of his latest research and just as his paper on geometry revolutionized the current views of space, finally freeing it from the Euclidean constraints, so his paper on number theory showed an entirely new and incredibly fruitful direction in which to head in pursuit of those unpredictable primes. It was not meant to be an attack on the Prime Number Theorem but to provide an entirely new way of counting primes and therefore of approximating $\pi(x)$ and did so by utilizing complex numbers and in particular the techniques of the new discipline of complex function theory. His approach was not rigorous but it scattered about the most fertile ideas as it rushed headlong through its eight pages and, using and refining these initiatives, two later mathematicians met with eventual success and finally provided the proof that had eluded so many for so long.

Legendre and Gauss had raised the issue of the nature of the prime counting function and with Gauss's involvement there is an inescapable feeling of déjà vu. It was he who had looked into the asymptotic statistical behaviour of almost all continued fractions and proposed a logarithmic solution involving a diminishing error term. The problem was not solved by him and it took a century before it was solved and then by two mathematicians, independently and nearly simultaneously, a result that brought some sort of order into a seemingly chaotic world. All of this is true of the Prime Number Theorem. Building on Riemann's ideas, de la Vallèe Poussin (whom we met on p. 113) and the Frenchman Jacques Hadamard (1865–1963) finally justified the word 'theorem' being used when in 1896 they showed that the relative error term in the approximation of $\pi(x)$ by $Li(x)$ was asymptotically zero. With all this profound mathematics around it is amusing to note that the proofs relied in part on the elementary trigonometric identity $3 + 4\cos\theta + \cos 2\theta = 2(1 + \cos\theta)^2 \geqslant 0$!

We will look more closely at Riemann's initiative in the final chapter but whatever the detail and with all the joy of success, it seemed unnatural that complex numbers were needed to prove a result about primes. Had a real-number proof escaped the scrutiny of the many mathematicians who had tried to find one? It seemed not as recently as 1932, since in that year the distinguished number theorist A. E. Ingham's much respected tract *The Distribution of Prime Numbers* was published, from which we gleaned that earlier proof of the 'Crucial Equivalence' of p. 184, and in the introduction he expressed the view:

> The solution (of the Prime Number Theorem) just outlined (that of de la Vallèe Poussin and Hadamard) may be held to be unsatisfactory in that it introduces ideas very remote from the original problem, and it is natural to ask for a proof of the Prime Number Theorem not depending on the theory of functions of a complex variable. To this we must reply that at present no such proof is known. We can indeed go further and say that it seems unlikely that a genuinely 'real variable' proof will be discovered, at any rate so long as the theory is founded on Euler's identity. For every known proof of the Prime Number Theorem is based on a certain property of the complex zeros of $\zeta(s)$, and this conversely is a simple consequence of the Prime Number Theorem itself. It seems clear therefore that this property must be used (explicitly or implicitly) in any proof based on $\zeta(s)$, and it is not easy to see how this is to be done if we take account only of real values of s.

It was no small matter, then, that in 1949 Atle Selberg (born 1917) published such a proof; indeed, it led to his award of the Fields Medal, which has played the role of the Nobel Prize in mathematics. Since that time other real-variable proofs have emerged, all termed 'elementary' and all fantastically difficult!

De la Vallèe Poussin was particularly interested in the size of the error term involved in the approximations of $\pi(x)$ and in an 1899 paper forever put to rest any doubts regarding primacy (!) among them. Confounding Legendre, and plenty of numeric evidence, he proved that 1 is asymptotically the optimal choice for a in the expression

$$\pi(x) = \frac{x}{\ln x - a} + \varepsilon_x.$$

(In 1962, Rosser and Schoenfeld showed that $x/(\ln x - 0.5) < \pi(x) < x/(\ln x - 1.5)$ for $x \geqslant 67$.) In the same paper he sounded the death knell for such estimates of $\pi(x)$ in that he proved, for large values of x, $Li(x)$ is better than any of them.

What has complex function theory to do with prime numbers? Just how accurate is the approximation of $Li(x)$ to $\pi(x)$? Simple enough question perhaps, but ones with very, very complicated answers.

The Riemann Initiative

The Zeta function is probably the most challenging and mysterious object of modern mathematics, in spite of its utter simplicity... The main interest comes from trying to improve the Prime Number Theorem, i.e. getting better estimates for the distribution of the prime numbers. The secret to the success is assumed to lie in proving a conjecture which Riemann stated in 1859 without much fanfare, and whose proof has since then become the single most desirable achievement for a mathematician.

M. C. Gutzwiller

16.1 COUNTING PRIMES THE RIEMANN WAY

In his paper Riemann considered another weighted prime counting function, which we will write as $\Pi(x)$, related to the harmonic series and defined by

$$\Pi(x) = \sum_{\substack{p^r < x, \\ p \text{ prime}}} \frac{1}{r},$$

which again reveals a bit more about itself if we look at a couple of examples:

$$\Pi(20) = \sum_{\substack{p^r < 20, \\ p \text{ prime}}} \frac{1}{r}$$

$$= \left(\frac{1}{1} + \frac{1}{2} + \frac{1}{3} + \frac{1}{4} \right)$$

$$+ \left(\frac{1}{1} + \frac{1}{2} \right) + \left(\frac{1}{1} \right) + \left(\frac{1}{1} \right) + \left(\frac{1}{1} \right) + \left(\frac{1}{1} \right) + \left(\frac{1}{1} \right) + \left(\frac{1}{1} \right),$$

where the bracketing is by the primes 2, 3, 5, ..., 19, and

$$\Pi(30) = \sum_{\substack{p^r < 30, \\ p \text{ prime}}} \frac{1}{r}$$

$$= \left(\frac{1}{1} + \frac{1}{2} + \frac{1}{3} + \frac{1}{4}\right) + \left(\frac{1}{1} + \frac{1}{2} + \frac{1}{3}\right) + \left(\frac{1}{1} + \frac{1}{2}\right)$$
$$+ \left(\frac{1}{1}\right) + \left(\frac{1}{1}\right) + \left(\frac{1}{1}\right) + \left(\frac{1}{1}\right) + \left(\frac{1}{1}\right) + \left(\frac{1}{1}\right) + \left(\frac{1}{1}\right),$$

where the bracketing is by the primes 2, 3, 5, ..., 29.

These can be rewritten as

$$\Pi(20) = \left(\frac{1}{1} + \frac{1}{1} + \frac{1}{1} + \frac{1}{1} + \frac{1}{1} + \frac{1}{1} + \frac{1}{1} + \frac{1}{1}\right)$$
$$+ \frac{1}{2}\left(\frac{1}{1} + \frac{1}{1}\right) + \frac{1}{3}\left(\frac{1}{1}\right) + \frac{1}{4}\left(\frac{1}{1}\right)$$

and

$$\Pi(30) = \left(\frac{1}{1} + \frac{1}{1} + \frac{1}{1} + \frac{1}{1} + \frac{1}{1} + \frac{1}{1} + \frac{1}{1} + \frac{1}{1} + \frac{1}{1} + \frac{1}{1}\right)$$
$$+ \frac{1}{2}\left(\frac{1}{1} + \frac{1}{1} + \frac{1}{1}\right) + \frac{1}{3}\left(\frac{1}{1} + \frac{1}{1}\right) + \frac{1}{4}\left(\frac{1}{1}\right).$$

The first bracket just counts the primes less than the number, the second those less than its square root, etc., to suggest in general that

$$\Pi(x) = \sum_{r=1}^{\infty} \frac{1}{r}\pi(x^{1/r}),$$

where, of course, there is in fact a finite number of terms.

The next step involved another of Gauss's students, August Möbius (1790–1868), who is most famously known for his one-sided band. He also produced a sophisticated 'changing the subject of a formula' technique known as Möbius Inversion to allow Riemann to arrive at the formula

$$\pi(x) = \sum_{r=1}^{\infty} \frac{\mu(r)}{r}\Pi(x^{1/r}),$$

where $\mu(r)$ is the Möbius function, which is somewhat esoterically defined by $\mu(1) = 1$ and

$$\mu(r) = \begin{cases} 0, & r \text{ has a repeated factor,} \\ 1, & r \text{ has an even number of prime factors,} \\ -1, & r \text{ has an odd number of prime factors.} \end{cases}$$

Taken out of context, this seems strange but the move is a standard number-theoretic one and not nearly as bizarre as a first impression suggests.

This is all very well, but all of this is of no use unless $\Pi(x)$ can be found by other means, and that other means was something of a favourite technique of Riemann's, and of a growing number of other contemporaries: the use of complex numbers and particularly complex function theory.

16.2 A NEW MATHEMATICAL TOOL

Two parts of the unique Parisian postal system are the 7th and 15th Arrondissements, and they are connected by more than adjacency: the 7th, apart from anything else, is home to Gustave Eiffel's tower, built as part of the World's Fair of 1889; the 15th to the Rue Cauchy. Each commemorates in its own way the contribution of Augustin Louis Cauchy (1789–1857), whose name appears on a plaque on the first stage of the tower, along with 71 other prominent French scientists. Whether one subscribes to the view that 'Cauchy was an admirable type of the true Catholic savant' or that he was possessed of 'self-righteous obstinacy and aggressive religious bigotry', he was a great mathematician and comparable to Euler in the volume of his mathematical output, which was as varied as it was profound, but unlike the mathematically flamboyant Euler, Cauchy was a rigorist and his contributions to the 19th century search for a firm foundation for mathematics were second to none. We are interested in his involvement in the development of complex function theory and many famous names appear in the list of those who advanced this important area of mathematics: Euler, Gauss, Riemann, d'Alembert, Laplace, Poisson, etc., but his stands above them all, although we will have need of only a small (but significant) part of the vast subject that it has become. In fact, to understand the impact of it on the study of prime numbers we will need three basic ideas from it: how to differentiate, how to integrate and the concept of analytic continuation. Differentiation is a very reasonable extension of the real case, with 'differentiable' equivalent to 'analytic'. Integration is more difficult (it always is) and requires the concept of integrating along a curve, or 'contour'. Analytic continuation is initially unbelievable. The technical details of complex differentiation and integration are approached in Appendix D; here we will simply put them to use, but first we need to define and appreciate analytic continuation

16.3 ANALYTIC CONTINUATION

The replacement of 'differentiable' by 'analytic' is more than semantic pedantry. Differentiation is essentially a limiting process and for a real function the limit can be approached from just two directions and must be independent of the direction chosen (which is why $f(x) = |x|$ is not differentiable at the origin).

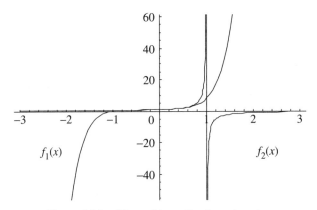

Figure 16.1. The problem of real continuation.

In the complex case there is an infinite number of possible directions, and again the answer must not depend on which of them is chosen. This makes great demands on the function and brings about strong results—one of which is analytic continuation. The process is probably best approached from the real case. For example, consider the functions

$$f_1(x) = 1 + x + x^2 + x^3 + \cdots \quad \text{and} \quad f_2(x) = \frac{1}{1-x} \quad \text{for } |x| < 1.$$

The theory of geometric series tells us that the first function converges only in the domain $|x| < 1$ and that the two functions are the same inside it. Plotting $f_1(x)$ to some number of terms and $f_2(x)$ on the same axes emphasizes that fact, and the difference between the two outside the interval (see Figure 16.1). There is not much sense in saying that the two are the same for $|x| > 1$, or that any of the infinite number of approximations to $f_1(x)$ will ever approach $f_2(x)$ in this region. Perhaps this all seems obvious, but replacing $x \in \mathbb{R}$ by $z \in \mathbb{C}$ changes everything since we have the following uniqueness theorem.

> *If, in some complex domain Δ, two analytic functions are defined and are equal at all points on a curve C lying inside Δ, they are equal throughout Δ.*

Let us pause to reflect on the enormity of what is being said. For example, suppose that two analytic functions are defined on the whole of $\mathbb{C}$ and are known to coincide just over the interval $[0, 1]$ on the real axis; then they must be equal everywhere else. Referring back to our example,

$$f_1(z) = 1 + z + z^2 + z^3 + \cdots = f_2(z) = \frac{1}{1-z} \quad \text{only for } |z| < 1$$

and is defined only in that circular region. Yet $f_2(z)$ is defined in all of $\mathbb{C}$, apart from $z = 1$, and so by the uniqueness theorem is *the* extension of $f_1(z)$. It is like a sleight-of-hand trick.

16.4 RIEMANN'S EXTENSION OF THE ZETA FUNCTION

Riemann's approach to the continuation of the Zeta function was to use contour integration and we deal with the details in Appendix E, but the result is that

$$\zeta(z) = \frac{\Gamma(1 - z)}{2\pi i} \oint_{u^-} \frac{u^{z-1}}{e^{-u} - 1} \, du$$

extends the definition of Zeta to all $z \neq 1$, for a particular contour u^-. We can see evidence of the Beautiful Relationship which we established on p. 60, with the real integral replaced by a particular contour integral.

16.5 ZETA'S FUNCTIONAL EQUATION

In a paper read in 1749 but not published until 1761, Euler suggested that the (real) Zeta functions satisfied the exotic functional relationship,

$$\zeta(1 - x) = \chi(x)\zeta(x),$$

where

$$\chi(x) = 2(2\pi)^{-x} \cos(\pi x/2)\Gamma(x).$$

He gave no proof but had verified the relationship to a point that, in his view, put the result beyond doubt. In the end the proof had to wait for Riemann and his complex generalization. By integrating around a second variable contour, which in the limit is the same as the original used to extend Zeta, the contour integral can be eliminated between two equations, leaving the above result, with real x generalized to complex z, and a form which conveniently reveals the important properties of the generalized Zeta function. Once again, the reader may wish to believe this or go to Appendix E for a proof.

16.6 THE ZEROS OF ZETA

If we look at a plot of the real, extended Zeta function (Figure 16.2), we can examine its behaviour for $x < 1$. The vertical asymptote at $x = 1$ is clear enough but on this scale the behaviour along the negative real axis is obscure and we need to zoom in a little, and doing so suggests that the function is zero at every negative, even integer.

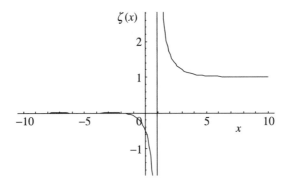

Figure 16.2. The real, extended Zeta function.

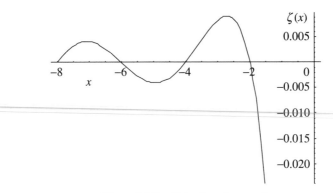

Figure 16.3. Behaviour for $x < 0$.

On p. 41 we saw that Euler had established Zeta's behaviour at positive, even integers in that

$$\zeta(2x) = \sum_{r=1}^{\infty} \frac{1}{r^{2x}} = (-1)^{x-1} \frac{(2\pi)^{2x}}{2(2x)!} B_{2x} \quad \text{for } x = 1, 2, 3, \ldots,$$

where the B_{2x} are the Bernoulli Numbers, but that the form of the Zeta function evaluated at odd positive integers (greater than 1, of course) remains a mystery to this day. In fact, the extended Zeta function is a little more compliant in that its exact form for all negative integers is known to be

$$\zeta(-x) = (-1)^x \frac{1}{x+1} B_{x+1} \quad \text{for } x = 0, 1, 2, \ldots,$$

which means that $\zeta(0) = -\frac{1}{2}$ and, since the other odd Bernoulli Numbers are all zero, it must be that the extended Zeta function is zero at negative even integers: these are called the trivial zeros—but there are others, which are not nearly so trivial.

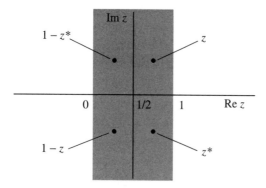

Figure 16.4. The symmetry of Zeta's non-trivial zeros.

The functional equation echoes this fact and also reveals a great deal more about the 'non-trivial' zeros of the Zeta function. If $z \in \{-2, -4, -6, \ldots\}$, $\cos(\pi z/2) \neq 0$ but $\Gamma(z)$ and therefore $\chi(z)$ is infinite, whereas $\zeta(1 - z)$ is finite; the only reconciliation is that $\zeta(z) = 0$ and we have those trivial zeros again. The Euler product form of $\zeta(z)$, valid only for $\mathrm{Re}(z) > 1$, clearly cannot be zero. For any zeros that do exist, the functional relationship tells us that if $\zeta(z) = 0$ and $\chi(z)$ is finite, then $\zeta(1 - z) = 0$. Therefore, there can be no other zeros for $\mathrm{Re}(z) < 0$, as such a zero would necessarily spawn another with its real part greater than 1. Riemann argued that there is an infinite number of these non-trivial zeros, which have a strong symmetry. In the interval $0 < \mathrm{Re}(z) < 1$, $\zeta(z)$ is a single-valued analytic function which is real when z is real; this is enough to mean that $(\zeta(z))^* = \zeta(z^*)$ (which is called the Schwartz Reflection Principle), and this means that $\zeta(z) = 0 \Leftrightarrow (\zeta(z))^* = 0 \Leftrightarrow \zeta(z^*) = 0$ (where z^* is the complex conjugate of z) and the symmetry becomes fourfold. It is hardly obvious, but no zeros lie on the line $\mathrm{Re}(z) = 1$ and using the functional equation none can lie on $\mathrm{Re}(z) = 0$. This seemingly minor detail is what de la Vallèe Poussin and Hadamard each established as an essential step in proving the Prime Number Theorem. In 1932 the eclectic, attractively eccentric American genius Norbert Wiener (1894–1964) showed that this result and the Prime Number Theorem are in fact entirely equivalent.

All non-trivial zeros of the Riemann Zeta function lie, then, symmetrically in the interval $0 < \mathrm{Re}(z) < 1$, which is known as the 'critical strip'; the shaded region in Figure 16.4, where $\zeta(z) = 0$.

It is of some interest to list the first few of these non-trivial zeros (with positive imaginary part), which we do in Table 16.1. The most striking feature is that the imaginary parts of each of the complex numbers is always 0.5: is this a representative selection? No one knows, but all available evidence suggests so and we will be addressing that critical matter soon; no one knows what those trailing dots suggest either—irrational, transcendental, etc.?

Table 16.1. Zeta's early non-trivial zeros.

14.134 725 141 734 693 790 457 251 983 562 470 270 784 257 115 699 243 · · · + 0.5i
21.022 039 638 771 554 992 628 479 593 896 902 777 334 340 524 902 781 · · · + 0.5i
25.010 857 580 145 688 763 213 790 992 562 821 818 659 549 672 557 996 · · · + 0.5i
30.424 876 125 859 513 210 311 897 530 584 091 320 181 560 023 715 440 · · · + 0.5i
32.935 061 587 739 189 690 662 368 964 074 903 488 812 715 603 517 039 · · · + 0.5i

16.7 THE EVALUATION OF $\Pi(x)$ AND $\pi(x)$

With the Zeta function analytically continued and with the symmetry of its zeros established, Riemann used contour integration again to develop a very striking expression for $\Pi(x)$ involving a very important infinite series,

$$\Pi(x) = Li(x) - \sum_{\rho} Li(x^\rho) - \ln 2 + \int_x^\infty \frac{du}{u(u^2 - 1)\ln u} \qquad x > 1. \quad (16.1)$$

The main things to notice about the formula are that $Li(x)$ appears together with a simple constant and another of those awkward integrals, which can be approximated to any accuracy for any given x; we also see an arresting series, which is summed over the infinity of zeros of the extended Zeta function. His argument was not fully rigorous and we will not attempt to repeat it here, but if we accept this mathematical alchemy for the moment, we can sum over any finite number of the zeros to arrive at an approximation of $\Pi(x^{1/r})$ for any x and the appropriate range of r, then use the expression

$$\pi(x) = \sum_{r=1}^\infty \frac{\mu(r)}{r} \Pi(x^{1/r})$$

to approximate $\pi(x)$; it seems a very tortuous route, but the diagrams in Figure 16.5 suggest that it is a very fruitful one

To make the mathematics sensible, it is necessary to define the step function $\pi(x)$ at the vertical step at each prime to be the midpoint of the rise; with this we can see that this process is able to take into account the local fluctuations in the behaviour of $\pi(x)$. In fact, if we look more closely at the contribution made by each of the Zeta function's non-trivial zeros we see that the kth zero contributes

$$Li(x^{\rho_k}) + Li(x^{\bar{\rho}_k})$$

to the sum and therefore

$$T_k(x) = \sum_{r=1}^\infty \frac{\mu(r)}{r}(Li(x^{\rho_k}) + Li(x^{\bar{\rho}_k}))$$

to $\pi(x)$. Some of the first few of these component functions are shown in Figure 16.6; notice the vertical scales—the early zeros contribute more significantly

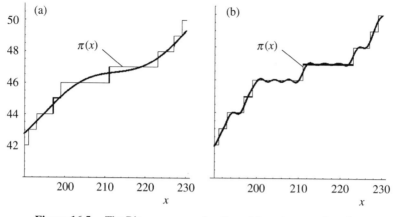

Figure 16.5. The Riemann approximation of the prime step function with (a) 10 terms and (b) 200 terms.

than those further on. The whole process is reminiscent of Fourier analysis and indeed the connection is profound: we are looking at the 'music of the primes'.

16.8 MISLEADING EVIDENCE

Looking back at Figure 15.8 on p. 176 shows that, at least up to 10^7, $Li(x) > \pi(x)$. This continues to be the case far, far beyond this value, in fact even today all available numeric evidence continues to point to $Li(x)$ being an overestimate of $\pi(x)$. Gauss always thought it to be true, so did Riemann, who at the end of his paper wrote,

> Indeed, in the comparison of $Li(x)$ with the number of prime numbers less than x, undertaken by Gauss and Goldschmidt and carried through up to x equals three million, this number has shown itself out to be, in the first hundred thousand, always less than $Li(x)$; in fact the difference grows, with many fluctuations, gradually with x.

$Li(x)$ seemed too big and Riemann suggested that it is in fact a closer approximation to a weighted sum of the $\pi(x)$ than it is to $\pi(x)$ alone; explicitly, that in his expression

$$\Pi(x) = \sum_{r=1}^{\infty} \frac{1}{r}\pi(x^{1/r}),$$

the $\Pi(x)$ might reasonably be replaced by $Li(x)$ itself to give

$$Li(x) \approx \pi(x) + \tfrac{1}{2}\pi(x^{1/2}) + \tfrac{1}{3}\pi(x^{1/3}) + \cdots$$

and by Möbius Inversion

$$\pi(x) \approx Li(x) - \tfrac{1}{2}Li(x^{1/2}) - \tfrac{1}{3}Li(x^{1/3}) - \cdots,$$

197

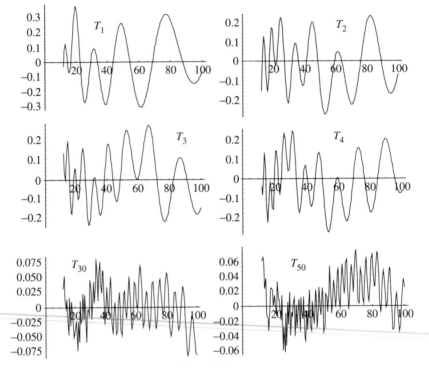

Figure 16.6.

an expression with dominant term $Li(x)$, but including an infinite series of refinements. And so we have a final approximating function

$$R(x) = \sum_{r=1}^{\infty} \frac{\mu(r)}{r} Li(x^{1/r}).$$

Figure 16.7 shows plots of this final approximation $R(x)$ with $\pi(x)$ and the difference between them fosters the hope that we do have an improvement for all x and for this to be true we clearly need that $Li(x) > \pi(x)$; unfortunately, it is not always so.

The leading quotation at the beginning of Chapter 7 was from the pen of Godfrey Harold Hardy, a complicated, modest, deeply gifted and influential number theorist, whom we have mentioned several times already. He is remembered for his own significant and individual contributions to mathematics but also those brought about by his collaboration with his great contemporary, John Edensor Littlewood (1885–1977). An incisive and elegant thumbnail picture of Littlewood appeared in a 1971/2 issue of the magazine, *Mathematical Spectrum*:

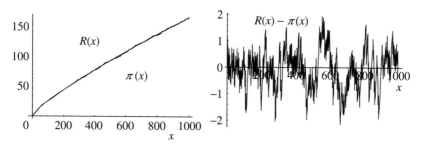

Figure 16.7. The Riemann estimate.

Fellow and Copley medallist of The Royal Society, honorary doctor
or member of many universities and academies, is the outstanding
mathematical analyst of his generation. Born in 1885, he has been
a Fellow of Trinity College Cambridge since 1908 and Rouse Ball
Professor of Mathematics from 1928 to 1950. Littlewood's papers
in analysis and Number Theory, of which over 100 were written
in collaboration with the late G. H. Hardy, have a striking power
which to mere mortals seems nothing short of miraculous.

Hardy agreed with '...knew of no one else who could command such a combination of insight, technique and power...'.

More romantically, Hardy and Littlewood are forever linked with the name
of the Indian genius Srinivasa Ramanujan (1887–1920), with the story of this
remarkable association told in *The Man Who Knew Infinity*. An example of
a typically extraordinary result of Ramanujan's is an exact formula for the
derivative of $\pi(x)$, with which we argued intuitively earlier. He proved that

$$\frac{\mathrm{d}\pi(x)}{\mathrm{d}x} = \frac{1}{x \ln x} \sum_{r=1}^{\infty} \frac{\mu(r)}{r} x^{1/r},$$

where the derivative of the step function is defined in terms of the usual limit.

All three took a profound interest in Number Theory in general and the Prime
Number Theorem in particular and it was Littlewood who, in 1914, proved that
eventually $\pi(x)$ will overtake $Li(x)$ and more, that the two functions will swap
in magnitude infinitely often from that point. Of course, this means that at these
values, $R(x)$ will not be the accurate approximation we would expect it to be. In
The Distribution of Prime Numbers, Ingham commented, 'This function ($R(x)$)
approximates $\pi(x)$ with astonishing accuracy for all values of x for which $\pi(x)$
has been calculated'. But he continues by remarking that, with Littlewood's
result, 'its superiority over the function $Li(x)$ is illusory' and that 'for special
values of x (as large as we please) the one approximation ($Li(x)$) will deviate
as widely as the other ($R(x)$) from the true value. On the bright side, he also
admits that 'on average' the first part of $R(x)$, $Li(x) - \frac{1}{2}Li(x^{1/2})$, will be a

199

better approximation to $\pi(x)$ than $Li(x)$ alone—at least, if something called the Riemann Hypothesis is true. The obvious question to ask is, what is the smallest value of x at which $\pi(x) > Li(x)$? To that question there remains no definite answer. In that same paper, Littlewood also proved that the asymptotic oscillations of the difference between the two functions are of the order of at least $Li(\sqrt{x})\ln\ln\ln x$, but he gave no explicit estimate of the whereabouts of that first sign change. Later, his student Stanley Skewes showed that it occurred before

$$10^{10^{10^{34}}},$$

which has become known as the 'Skewes Number', and which at the time was the biggest 'useful' number ever defined ('Graham's Number', from the world of combinatorics, now dwarfs it). As of 2000, Carter Bays and Richard Hudson have improved the bound by showing that the first change of sign occurs before a mere $1.398\,22 \times 10^{316}$, a number still far beyond present-day computational reach.

16.9 THE VON MANGOLDT EXPLICIT FORMULA — AND HOW IT IS USED TO PROVE THE PRIME NUMBER THEOREM

It was left to others to recast Riemann's thoughts with the severity that mathematics ultimately demands and in this case the most notable contributor was Von Mangoldt, who provided a rigorous proof of Riemann's Equation (16.1) but who also established a similar expression for the ψ function described on p. 196, and which has overtaken $\Pi(x)$ in the study of the Prime Number Theorem. It is this form that we will look at in some detail.

We have that the complex form of the Euler identity is

$$\zeta(z) = \prod_{p \text{ prime}} \frac{1}{1 - p^{-z}},$$

and is valid for $\mathrm{Re}(z) > 1$, and so

$$\ln\zeta(z) = \ln\prod_{p \text{ prime}} \frac{1}{1 - p^{-z}} = -\sum_{p \text{ prime}} \ln(1 - p^{-z})$$
$$= -\sum_{p \text{ prime}} \ln(1 - e^{-z\ln p}).$$

Differentiating with respect to z then gives

$$\frac{\zeta'(z)}{\zeta(z)} = -\sum_{p \text{ prime}} \frac{e^{-z\ln p}\ln p}{1 - e^{-z\ln p}} = -\sum_{p \text{ prime}} \frac{p^{-z}\ln p}{1 - p^{-z}} = -\sum_{\substack{p \text{ prime}\\ r=1}}^{\infty} \frac{\ln p}{p^{rz}}. \quad (16.2)$$

The last expression uses the sum of an infinite geometric series. We will use the $\psi(x) \sim x$ form of the Prime Number Theorem and recalling the definition $\psi(x) = \sum_{p^r \leq x} \ln p$, we will naturally seek to extract the logarithmic part from the sum on the right-hand side of Equation (16.2), which can be done using the contour integral device

$$\frac{1}{2\pi i} \int_{c-i\infty}^{c-i\infty} \frac{y^z}{z}\, dz = \begin{cases} 0, & 0 < y < 1, \\ \frac{1}{2}, & y = 1, \\ 1, & y > 1, \end{cases}$$

where c is a convenient real number. Once again for those who are aware of them, the techniques of Fourier analysis are familiar.

Multiplying both sides of the expression (16.2) by x^z/z and rearranging gives

$$\frac{x^z}{z} \sum_{\substack{p \text{ prime} \\ r=1}}^{\infty} \frac{\ln p}{p^{rz}} = \sum_{\substack{p \text{ prime} \\ r=1}}^{\infty} \left(\frac{x}{p^r}\right)^z \frac{\ln p}{z} = -\frac{\zeta'(z)}{\zeta(z)} \frac{x^z}{z},$$

and so integrating both sides along the contour gives

$$\frac{1}{2\pi i} \int_{c-i\infty}^{c-i\infty} \sum_{\substack{p \text{ prime} \\ r=1}}^{\infty} \left(\frac{x}{p^r}\right)^z \frac{\ln p}{z}\, dz = \frac{1}{2\pi i} \int_{c-i\infty}^{c-i\infty} -\frac{\zeta'(z)}{\zeta(z)} \frac{x^z}{z}\, dz,$$

$$\sum_{\substack{p \text{ prime} \\ r=1}}^{\infty} \ln p \frac{1}{2\pi i} \int_{c-i\infty}^{c-i\infty} \left(\frac{x}{p^r}\right)^z \frac{1}{z}\, dz = \frac{1}{2\pi i} \int_{c-i\infty}^{c-i\infty} -\frac{\zeta'(z)}{\zeta(z)} \frac{x^z}{z}\, dz.$$

Now take $y = x/p^r$ to get

$$\sum_{\substack{p \text{ prime} \\ r=1}}^{\infty} \ln p \frac{1}{2\pi i} \int_{c-i\infty}^{c-i\infty} \frac{y^z}{z}\, dz = \frac{1}{2\pi i} \int_{c-i\infty}^{c-i\infty} -\frac{\zeta'(z)}{\zeta(z)} \frac{x^z}{z}\, dz$$

and

$$\psi(x) = \sum_{p^r < x}^{\infty} \ln p = \frac{1}{2\pi i} \int_{c-i\infty}^{c-i\infty} -\frac{\zeta'(z)}{\zeta(z)} \frac{x^z}{z}\, dz$$

since $p^r > x$ would mean $y < 1$ and the integral contribution 0; x must not be the power of a prime. The remaining contour integral is evaluated using the theory of residues, all of which have to be added together to arrive at the answer. The integral is best thought of divided into four different categories of residue, as in Table 16.2.

Table 16.2. The four types of residue.

Singularity	Cause	Residue
0	$\dfrac{x^z}{z}$	$\dfrac{\zeta'(0)}{\zeta(0)} = \ln 2\pi$
1	Pole of ζ	$-\dfrac{x^1}{1} = -x$
$-2, -4, -6, -8, \ldots$	Trivial zeros of ζ	$\frac{1}{2}x^{-2}, \frac{1}{4}x^{-4}, \frac{1}{6}x^{-6}, \frac{1}{8}x^{-8}, \ldots$
ρ	Non-trivial zeros of ζ	$\dfrac{x^\rho}{\rho}$

Yet again the Taylor series for ln appears, this time as

$$\tfrac{1}{2}x^{-2} + \tfrac{1}{4}x^{-4} + \tfrac{1}{6}x^{-6} + \tfrac{1}{8}x^{-8} + \cdots = \tfrac{1}{2}\ln(1 - x^{-2})$$

and we therefore have

$$\psi(x) = x - \ln(2\pi) - \tfrac{1}{2}\ln(1 - x^{-2}) - \sum_{\zeta(\rho)=0} \frac{x^\rho}{\rho},$$

where the sum is over the non-trivial zeros, which is the equivalent of Riemann's expression for $\Pi(x)$. It is known as the Von Mangoldt explicit formula and has to be the most important in the whole of analytic number theory. At first it looks contradictory to have a real function on the left in part made up from an infinite sum of complex numbers, but the roots do occur in conjugate pairs, which makes the terms, taken in such pairs, real.

Now we can see the connection between the Prime Number Theorem and ζ's zeros. If we write $\rho = u + iv$, then $|x^\rho| = x^u$ and $u < 1$ would mean that, as $x \to \infty$, each error term in the series is of order less than x and this would mean (with a bit more mathematical rigour) that $\psi(x)/x \to 1$, as required. That is, the real part of the non-trivial zeros of the extended Zeta function being less than 1 would imply the Prime Number Theorem and, as we have said, it was this fact that de la Vallèe Poussin and Hadamard independently established.

16.10 THE RIEMANN HYPOTHESIS

In his paper, Riemann defined a function ξ, related to ζ, by

$$\xi(w) = \pi^{-z/2}(z - 1)\Gamma(\tfrac{1}{2}z + 1)\zeta(z),$$

where $z = \tfrac{1}{2} + iw$. Why? Really, because it is easier to handle than $\zeta(z)$. The $(z - 1)$ eliminates the problem with $\zeta(z)$ at $z = 1$ (recall from p. 41 that $(z - 1)\zeta(z) \to 1$ as $z \to 1$) and so ξ is analytic in the whole complex plane, it

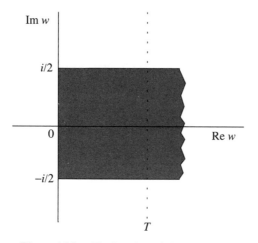

Figure 16.8. The location of ξ's early zeros.

is also not hard to check that $\xi(z) = \xi(1-z)$ and from the definition it's clear that the set of zeros of ξ is the same as the set of zeros of ζ. What is more, the fact that all of the non-trivial zeros of ζ lie in $0 < \mathrm{Re}(z) < 1$ means that if we write $\xi(w) = \xi(u+iv) = 0$, then $\zeta(z) = 0$, where $z = (\frac{1}{2} - v) + iu$ and so we must have that $0 < \frac{1}{2} - v < 1$ and so $-\frac{1}{2} < v < \frac{1}{2}$; that is, the zeros of ξ must have imaginary parts lying between $-\frac{1}{2}$ and $\frac{1}{2}$. Using the symmetry of the zeros of ζ we need only consider those which have a positive imaginary part, making $u > 0$ and therefore $\mathrm{Re}(w) > 0$. This results in the region in Figure 16.8.

Riemann argued (again vaguely) that about

$$\frac{T}{2\pi} \ln \frac{T}{2\pi} - \frac{T}{2\pi}$$

of the zeros lie in such a rectangle and as a test he calculated the real zeros, to find that the number closely agreed with the counting function, which left little space for any others. In his own words, 'One now finds indeed approximately this number of real roots within these limits, and it is very probable that all roots are real'. If this is the case, the real part of the zeros of ζ must be $\frac{1}{2}$. Continuing, he remarked, 'Certainly one would wish for a stricter proof here; I have meanwhile temporarily put aside the search for this after some fleeting futile attempts, as it appears unnecessary for the next objective of my investigation.' Which brings us to the vaunted

---

Riemann Hypothesis

The non-trivial zeros of the Riemann Zeta function
all have real part one-half

203

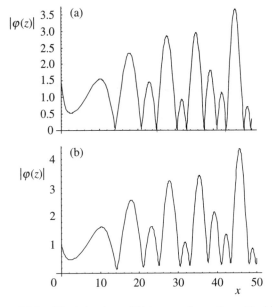

Figure 16.9. The behaviour of Zeta on and near the critical line.
(a) $z = \frac{1}{2} + xi$; (b) $z = \frac{1}{3} + xi$.

In terms of Figure 16.4 on p. 195, the zeros all lie on the line of symmetry rather than in any other part of the critical region.

The two plots in Figure 16.9 show the early behaviour of the function $|\zeta(z)|$ for points on the critical line $z = \frac{1}{2} + xi$ and for points on the parallel line $z = \frac{1}{3} + xi$. They also show that $\zeta(z)$ has plenty of zeros at the start of the vertical line $\text{Re}(z) = \frac{1}{2}$ but none such on $\text{Re}(z) = \frac{1}{3}$, although it can come perilously close, as can be seen near the point $\frac{1}{3} + 14i$.

Plotting $1/|\zeta(z)|$ in Figure 16.10 gives another revealing glimpse of the non-trivial zeros, which appear as spikes along the line $\text{Re}(z) = \frac{1}{2}$. The trivial zeros bring about the 'mountain' on the left.

In passing, Hadamard established the very satisfying form

$$\xi(w) = -e^{-Az} \prod_{\zeta(\rho)=0} \left(1 - \frac{z}{\rho}\right) e^{z/\rho},$$

where $A = -\frac{1}{2}\gamma - 1 + \frac{1}{2}\ln 4\pi$.

16.11 WHY IS THE RIEMANN HYPOTHESIS IMPORTANT?

The Riemann Hypothesis states that all non-trivial roots of the Zeta function have real part $\frac{1}{2}$, a far stronger condition than the one required for the proof of the

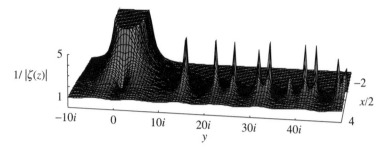

Figure 16.10. A three-dimensional view of Zeta's early zeros.

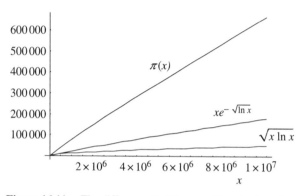

Figure 16.11. The difference the Riemann Hypothesis makes.

Prime Number Theorem, which merely requires that none have real part 1. The immediate importance of the conjecture is in the measurement of the size of the error involved in the approximation of $\pi(x)$ by $Li(x)$, but it strikes much deeper and into the greatest depths of mathematics, with the error involved in many important asymptotic formulae also governed by it: for example, the weaker form of the Goldbach Conjecture, which states that every odd number is the sum of three primes, is implied by it. Fields Medallist, Enrico Bombieri, has said that 'The failure of the Riemann Hypothesis would create havoc in the distribution of prime numbers'. Since the Riemann Hypothesis in involved with the size of the error in approximating $\psi(x)$ by x, it therefore is involved with the error in approximating $\pi(x)$ by $Li(x)$. To be exact, in 1901 von Koch proved that, if the Riemann Hypothesis is true, the known estimate $\pi(x) = Li(x) + O(xe^{-c\sqrt{\ln x}})$ would become $\pi(x) = Li(x) + O(\sqrt{x}\ln x)$, which Bombieri has commented would be hard to significantly improve on, given Littlewood's result that the degree of oscillation of $\pi(x) - Li(x)$ is asymptotically of the order $Li(\sqrt{x}) \times \ln\ln\ln x$. Figure 16.11 gives some sort of idea of the difference between the size of the errors with and without the Riemann Hypothesis.

Proving the Riemann Hypothesis has subsequently become the greatest prob-

lem in mathematics but has largely resisted attempts by some of the best mathematicians of the 20th century to gain significant headway with it.

16.12 REAL ALTERNATIVES

The uniqueness theorem allows us the freedom to extend Zeta's definition in
any way we please and various methods have been used to do just that, including the use of Euler–Maclaurin summation; of course, Riemann used contour
integration, which reveals a great deal about the nature of the extended function. Another approach uses the generalized alternating harmonic series, the
'alternating Zeta function', defined by

$$\zeta_a(z) = \sum_{r=1}^{\infty} \frac{(-1)^r}{r^z},$$

which converges in the bigger region $\operatorname{Re}(z) > 0$.
    We can write this as

$$\zeta_a(z) = 1 - \frac{1}{2^z} + \frac{1}{3^z} - \frac{1}{4^z} + \frac{1}{5^z} + \cdots$$

$$= 1 + \frac{1}{2^z} + \frac{1}{3^z} + \frac{1}{4^z} + \frac{1}{5^z} + \cdots - 2\left(\frac{1}{2^z} + \frac{1}{4^z} + \frac{1}{6^z} + \cdots\right)$$

$$= 1 + \frac{1}{2^z} + \frac{1}{3^z} + \frac{1}{4^z} + \frac{1}{5^z} + \cdots - \frac{2}{2^z}\left(1 + \frac{1}{2^z} + \frac{1}{3^z} + \cdots\right)$$

and so

$$\zeta_a(z) = \zeta(z) - \frac{1}{2^{z-1}}\zeta(z),$$

which makes

$$\zeta(z) = \frac{\zeta_a(z)}{1 - 2^{1-z}} = \frac{1}{1 - 2^{1-z}} \sum_{r=1}^{\infty} \frac{(-1)^r}{r^z}, \tag{16.3}$$

defined for $\operatorname{Re}(z) > 0$.
    The extension is made complete using yet another technique of Euler's,
'Euler's series transformation', and this results in

$$\zeta(z) = \frac{1}{1 - 2^{1-z}} \sum_{r=0}^{\infty} \frac{1}{2^{r+1}} \sum_{k=0}^{r} (-1)^k \binom{r}{k} (k+1)^{-z}$$

for $z \neq 1$.
    It seems light years away from the contour integral form, but remember that
uniqueness theorem for analytic extension! We can use the extension (16.3) to

give a tantalizingly simple reformulation of the Riemann Hypothesis without complex numbers appearing at all. Using standard methods,

$$r^z = r^{a+ib} = r^a r^{ib} = r^a e^{ib\ln r} = r^a(\cos(b\ln r) + i\sin(b\ln r))$$

and so

$$\frac{1}{r^z} = \frac{1}{r^a}(\cos(b\ln r) - i\sin(b\ln r)),$$

which means that

$$\zeta(z) = 0 \Leftrightarrow \sum_{r=1}^{\infty} \frac{(-1)^r}{r^z} = 0$$

$$\Leftrightarrow \sum_{r=1}^{\infty} \frac{(-1)^r}{r^a}(\cos(b\ln r) - i\sin(b\ln r)) = 0.$$

Equating real and imaginary parts brings us to the very tempting reformulation:

> If $\displaystyle\sum_{r=1}^{\infty} \frac{(-1)^r}{r^a}\cos(b\ln r) = 0$ and $\displaystyle\sum_{r=1}^{\infty} \frac{(-1)^r}{r^a}\sin(b\ln r) = 0$
>
> for some pair of real numbers a and b, then $a = \frac{1}{2}$.

The reader may wish to check this using the early zeros given in Table 16.1 on p. 196. It seems extraordinary that the most famous unsolved problem in the whole of mathematics can be phrased so that it involves the simplest of mathematical ideas: summation, trigonometry, logarithms and of course, if the conjecture is true, Christof Rudolff's $\sqrt{}$ sign. It all sounds so easy to become the most famous name in the mathematical world!

There are other, equivalent real formulations of the Riemann Hypothesis. For example, asymptotically the exact values of the integers $\lfloor Li(x) \rfloor$ and $\pi(x)$ must agree on 'about' half of their digits. Also, with $\sigma(n)$ the sum of the divisors of n, that $\sigma(n) < e^\gamma n \ln\ln n$ for $n \geqslant 5041$ or that $\sigma(n) \leqslant H_n + e^{H_n} \ln H_n$ for $n \geqslant 1$, with equality only for $n = 1$. We will content ourselves with a detailed look at one more celebrated reformulation.

16.13 A BACK ROUTE TO IMMORTALITY—PARTLY CLOSED

Any integer can be written as the product of a square and a square-free component and in Chapter 3 we saw this simple fact put to significant use by Erdos. Of course, any particular integer might be factored as a combination of square and square-free, for example, $2^3 \times 3^5 \times 7 \times 11^2 = (2 \times 3^2 \times 11)^2(2 \times 3 \times 7)$, or it could be a perfect square, $3^6 \times 5^4 \times 13^2 = (3^3 \times 5^2 \times 13)^2$, or it could be entirely square free, with the primes appearing just once, for example, $2 \times 5 \times 13 \times 17$.

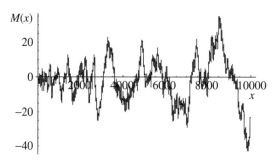

Figure 16.12. Mertens function.

The Möbius function μ, mentioned earlier, is used to discriminate between the types of factorization that are possible. Recall its definition:

$$\mu(r) = \begin{cases} 0, & r \text{ has a repeated factor,} \\ 1, & r \text{ has an even number of prime factors,} \\ -1, & r \text{ has an odd number of prime factors.} \end{cases}$$

Now suppose that we consider all square-free integers. It is reasonable to suppose that the Almighty has divided them pretty equally so that μ will take its values of $+1$ and -1 equally often (in fact, it can be shown that $P(\mu(r) = 1) = P(\mu(r) = -1) = 3/\pi^2$, and therefore $P(\mu(r) = 0) = 1 - 6/\pi^2$, giving a final appearance of that ubiquitous number). Having said this, we would expect some fluctuation in the count as we move along the list of integers—just as we have expected fluctuations in the accuracy of $Li(x)$ approximating $\pi(x)$ or any other asymptotic approximation. But how big would we expect those fluctuations to be? The size of them is measured by the absolute value of the Mertens function $M(x) = \sum_{r \leqslant x} \mu(r)$, shown in Figure 16.12.

It is clearly erratic but even so, in 1885 Thomas Stieltjes (1856–1894), 'the father of the analytic theory of continued fractions', claimed in a letter to his frequent correspondent Charles Hermite (1822–1901) that $M(x)x^{-1/2}$ stays within two fixed bounds, no matter how large x may be; he added (in parenthesis) that the bounds could probably be taken to be $+1$ and -1. In saying this, he was suggesting that $|M(x)| < \sqrt{x}$. In 1897, Mertens published a paper containing a table 50 pages long giving values of $\mu(r)$ and $M(r)$ for r up to $10\,000$ and based on this evidence claimed that Stieltjes stronger estimate was 'very probable' and so $|M(x)| < \sqrt{x}, x > 1$, passed into mathematics as the 'Mertens Conjecture'. In a series of papers over the turn of the century, von Sterneck published values of $M(r)$ for r up to $1\,000\,000$ and on that evidence conjectured the stronger $|M(x)| < 0.5\sqrt{x}, x > 200$.

Stieltjes's proof never appeared because the assertion is wrong, which means that the von Sterneck assertion is wrong too, and even the weaker forms, with larger bounds, might be doomed to failure also. It took until 1963 to disprove the stronger form, when Gerhard Neubauer found that with $x = 7\,725\,038\,629$, the 0.5 but not the 1 boundary is broken. Not until 1985 was the original conjecture dispatched, when A. M. Odlyzko and H. J. J. te Riele proved that eventually the positive and the negative barriers are broken. (With that erratic behaviour, it is hardly surprising that they formulated their result in terms of the ideas on p. 113; to be exact, they showed that $\lim \sup_{x \to \infty} M(x)x^{-1/2} > 1.06$ and $\lim \inf_{x \to \infty} M(x)x^{-1/2} < -1.009$.) Their proof was one of existence and as such provided no estimate, let alone value, for such an x; in the same year Janos Pintz proved that the first counterexample must be less than 3.21×10^{64}—big, but bear in mind the Skewes and Graham numbers!

This all seems a shame, with numeric evidence once again leading intuition astray; a few million, a few billion, a few trillion... do not mean much here; in number theory, big really can mean BIG!

What has it to do with the Riemann Hypothesis? Its truth would have implied it. In fact, the truth of $|M(x)| < C\sqrt{x}$ for any constant C would imply it—and that remains an open question; small wonder that the conjecture has attracted the attention that has led to two of its forms being disproved.

The Zeta function is intimately related to the Möbius function in that

$$\frac{1}{\zeta(z)} = \sum_{r=1}^{\infty} \frac{\mu(r)}{r^z} \quad \text{for } \mathrm{Re}(z) > 1.$$

We will not prove this fact, but it another standard result of number theory. With one last look at complex function theory and with this result at our disposal, we can see that tantalizing connection, given that we define $M(0) = 0$:

$$\frac{1}{\zeta(z)} = \sum_{r=1}^{\infty} \frac{\mu(r)}{r^z} = \sum_{r=1}^{\infty} \frac{M(r) - M(r-1)}{r^z}$$

$$= \sum_{r=1}^{\infty} \frac{M(r)}{r^z} - \sum_{r=1}^{\infty} \frac{M(r-1)}{r^z} = \sum_{r=1}^{\infty} \frac{M(r)}{r^z} - \sum_{r=1}^{\infty} \frac{M(r)}{(r+1)^z}$$

$$= \sum_{r=1}^{\infty} M(r) \left\{ \frac{1}{r^z} - \frac{1}{(r+1)^z} \right\} = \sum_{r=1}^{\infty} M(r) \int_r^{r+1} \frac{z}{x^{z+1}} \, dx$$

$$= z \sum_{r=1}^{\infty} \int_r^{r+1} \frac{M(x)}{x^{z+1}} \, dx = z \int_1^{\infty} \frac{M(x)}{x^{z+1}} \, dx$$

since $M(x)$ is constant on each interval $[r, r+1)$.

If the Mertens conjecture is true, then

$$\left| \frac{M(x)}{x^{z+1}} \right| < \left| \frac{C\sqrt{x}}{x^{z+1}} \right| = \frac{C}{\sqrt{x}} \left| \frac{1}{x^z} \right| = \frac{C}{\sqrt{x}} \frac{1}{x^{\mathrm{Re}(z)}} = \frac{C}{x^{\mathrm{Re}(z)+1/2}}.$$

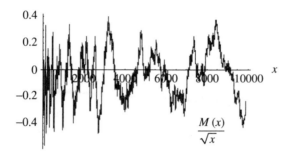

Figure 16.13. Early evidence for Stieltjes conjecture.

The last integral would converge provided that $\text{Re}(z) + \frac{1}{2} > 1$, which means that $\text{Re}(z) > \frac{1}{2}$. If this is so, it would define a function analytic in $\text{Re}(z) > \frac{1}{2}$, which would give an analytic continuation of $1/\zeta(z)$ from $\text{Re}(z) > 1$ in the original formula to $\text{Re}(z) > \frac{1}{2}$ (that sleight of hand again). This would mean that $1/\zeta(z)$ is defined for $\text{Re}(z) > \frac{1}{2}$ (and therefore that $\zeta(z)$ can have no zeros there); by symmetry, none could exist in $\text{Re}(z) < \frac{1}{2}$, so they all must lie on $\text{Re}(z) = \frac{1}{2}$ and that of course is the Riemann Hypothesis!

16.14 INCENTIVES, OLD AND NEW

Mathematical Problems

Lecture delivered before the International Congress of
Mathematicians at Paris in 1900

By Professor David Hilbert

Who of us would not be glad to lift the veil behind which the future
lies hidden; to cast a glance at the next advances of our science and
at the secrets of its development during future centuries? What
particular goals will there be toward which the leading mathematical spirits of coming generations will strive? What new methods
and new facts in the wide and rich field of mathematical thought
will the new centuries disclose? History teaches the continuity of
the development of science. We know that every age has its own
problems, which the following age either solves or casts aside as
profitless and replaces by new ones. If we would obtain an idea
of the probable development of mathematical knowledge in the
immediate future, we must let the unsettled questions pass before
our minds and look over the problems which the science of today
sets and whose solution we expect from the future. To such a review
of problems the present day, lying at the meeting of the centuries,

seems to me well adapted. For the close of a great epoch not only invites us to look back into the past but also directs our thoughts to the unknown future. The deep significance of certain problems for the advance of mathematical science in general and the important role which they play in the work of the individual investigator are not to be denied. As long as a branch of science offers an abundance of problems, so long is it alive; a lack of problems foreshadows extinction or the cessation of independent development. Just as every human undertaking pursues certain objects, so also mathematical research requires its problems. It is by the solution of problems that the investigator tests the temper of his steel; he finds new methods and new outlooks, and gains a wider and freer horizon. It is difficult and often impossible to judge the value of a problem correctly in advance; for the final award depends upon the gain which science obtains from the problem. Nevertheless we can ask whether there are general criteria which mark a good mathematical problem. An old French mathematician said: 'A mathematical theory is not to be considered complete until you have made it so clear that you can explain it to the first man whom you meet on the street.' This clearness and ease of comprehension, here insisted on for a mathematical theory, I should still more demand for a mathematical problem if it is to be perfect; for what is clear and easily comprehended attracts, the complicated repels us. Moreover a mathematical problem should be difficult in order to entice us, yet not completely inaccessible, lest it mock at our efforts. It should be to us a guide post on the mazy paths to hidden truths, and ultimately a reminder of our pleasure in the successful solution.

On 8 August 1900 David Hilbert (1862–1943) rose to a lecturn in the Sorbonne to give what is probably the most famous lecture ever delivered by a mathematician (although Andrew Wiles's series of lectures, in which he established a form of the Tanayama–Shimura conjecture and in particular Fermat's Last Theorem—admittedly with a later corrected error—might vie for equal renown). Hilbert, even with the formidable competition of the likes of Felix Klein and Henri Poincaré, was the most acclaimed mathematician of his day, described by one of his students (a future Nobel Laureate) by the words, '. . . lives in my memory as perhaps the greatest genius I ever laid eyes on.' He had been invited to give one of the major addresses at the second International Congress of Mathematicians and he chose to use the opportunity to chart a course for 20th-century mathematics, in part by posing a series of 23 problems, the investigation or solution of which would in his view lead the way to mathematical progress. The address opened with the lines above and continued by focusing on 10 of the problems; there was no apparent order to his list but on it, and one discussed in the address, was problem number eight.

8. Problems of prime numbers

Essential progress in the theory of the distribution of prime numbers has lately been made by Hadamard, de la Vallée Poussin, Von Mangoldt and others. For the complete solution, however, of the problems set us by Riemann's paper 'Ueber die Anzahl der Primzahlen unter einer gegebenen Grösse', it still remains to prove the correctness of an exceedingly important statement of Riemann, viz., that the zero points of the function $\zeta(s)$ defined by the series

$$\zeta(s) = 1 + \frac{1}{2^s} + \frac{1}{3^s} + \frac{1}{4^s} + \cdots$$

all have the real part 1/2, except the well-known negative integral real zeros. As soon as this proof has been successfully established, the next problem would consist in testing more exactly Riemann's infinite series for the number of primes below a given number and, especially, to decide whether the difference between the number of primes below a number x and the integral logarithm of x does in fact become infinite of an order not greater than 1/2 in x. Further, we should determine whether the occasional condensation of prime numbers which has been noticed in counting primes is really due to those terms of Riemann's formula which depend upon the first complex zeros of the function $\zeta(s)$.

Hilbert's gigantic standing gave huge impetus in the mathematical world to address the problems in the list—a reputation could be made by success in any of them. Those who did meet with success, or who contributed significantly to success were to become known as members of the 'honours class' of mathematicians. Of the 23 problems, 8 were of a purely investigative nature and 12 of the remaining 15 have been completely resolved. Only problem number 8 preserves its mystery almost completely and a century later it remains, in a practical sense, untouched.

In 1998 the Fields Medallist, Steven Smale, put forward his own list of 18 problems in the same spirit as Hilbert and on 13 February 2002 the solution of the 14th on the list was published by W. Tucker. So far this is the only one of them to be solved, and number one on the list is the Riemann Hypothesis.

With the dawn of another millennium, a new incentive has been provided by the Clay Mathematics Institute in that they have offered one million dollars each for the solution of seven open questions, one of which is the Riemann Hypothesis.

16.15 PROGRESS

There has, of course, been progress. In 1914 Hardy wrote the paper 'Sur les zeros de la fonction $\zeta(z)$ de Riemann' in which he showed that an infinite number of the non-trivial zeros lie on the critical line $\mathrm{Re}(z) = \frac{1}{2}$. In 1921, he and Littlewood together proved the far stronger result that, for some positive constant A, $\zeta(\frac{1}{2} + iy)$ has at least AY zeros in each interval $-Y \leqslant y \leqslant Y$. Selberg, in 1942, improved Hardy's original result to show that a positive proportion of all the non-trivial zeros lie on the critical line. (This is a subtle but important distinction. For example, $\mathbb{Z}$ is infinite but the precise 'measure' of its size compared with $\mathbb{R}$ is 0.) Conrey improved this in 1989, showing that at least 40% of the zeros lie on the line. The width of the critical region has been squeezed, but not to zero, which you may think is pretty convincing evidence, but recall the two conjectures mentioned earlier; Littlewood was far from convinced: he conjectured that the Riemann Hypothesis is false!

Since there is known to be an infinite number of non-trivial zeros with no discernible pattern to them, enumerating them is not an option—other than to hope to find one not on the critical line. To this end, in 1903 J.-P. Gram used Euler–Maclaurin summation to prove that the conjecture is true for a height of 50, that is, for $\mathrm{Im}(z) < 50$, but Euler–Maclaurin summation has long been superseded by a clever technique on which we will touch lightly.

Recall that $\xi(z) = \xi(1 - z)$ and also that the function is analytic and real for real z. This means that we can use the Schwartz reflection formula again and, in particular, we have

$$(\xi(\tfrac{1}{2} + it))^* = \xi((\tfrac{1}{2} + it)^*) = \xi(\tfrac{1}{2} - it) = \xi(1 - (\tfrac{1}{2} + it)) = \xi(\tfrac{1}{2} + it)$$

and the only complex numbers equal to their own conjugates are real. We have that ξ is real on the critical line and so to look for a zero on the line is to look for a change in sign of the ξ function. (The precise method for achieving this is technical and uses something known as Gram's Law.) Now all we need to do is to provide an accurate count of how many zeros exist up to a certain height and compare that number with the count of the number of zeros on the critical line: any discrepancy proves the hypothesis false. And this takes us to our final genius. Recall that Ada Lovelace thought an appropriate task for Babbage's Calculating Engine was the evaluation of the Bernoulli Numbers; the eccentric and pitifully treated British genius Alan Turing (1912–1954) felt that locating zeros of the Riemann Zeta function was an appropriate task for the Calculating Engine's successor—the electronic computer—the intellectual form of which he conceived. Turing is most generally remembered for his immense contributions to the breaking of the German military Enigma Code at Bletchely Park, England, in World War II; the gripping story of 'Ultra' has been told by many now that it is not shrouded by the Official Secrets Act, the intellectual 'cream of the cream' acting in unison to achieve what was though to be impossible. Even in that most

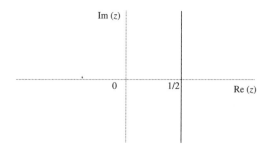

Figure 16.14. Proof without words: the Riemann Hypothesis.

rarefied atmosphere Turing, the 'Prof', was special; his own story has been told in Andrew Hodges's *Alan Turing; the Enigma* and Jon Agar's *Turing and the Universal Machine* (among others), and we will merely give fleeting mention of one of his many brilliant ideas.

In 1948 he was at Manchester university, belatedly joining the team who constructed the first electronic, stored-program computer and it was from here that he put forward his seminal ideas on machine intelligence. By 1951 the machine had graduated to the 'Blue Pig' or MUC, the Manchester University Computer, a massive collection of wiring and valves (concealed in metal cupboards) which was set to many tasks from singing and producing doggerel to testing for the zeros of $\xi(z)$. At night, when it had no other work, Turing would set it to work widening the search and using a formula devised by him (and still used) to provide an accurate count of the number of zeros up to a given height. The search was futile and the evidence continues to build far beyond the reach of the Blue Pig that the two counts match; now it is known that 59 974 310 000 zeros lie on the line—and of course none have been found off it!

We have mentioned G. H. Hardy several times before and he was one of the outstanding mathematicians of his time, making many significant contributions to number theory. In his immensely impressive mathematical trophy cabinet there was a vast gap waiting to be filled by a proof of the Riemann Hypothesis, a gap that remained empty, of course, and we can gain some small insight into the man and his view of the Riemann Hypothesis with these three anecdotes.

- A list of his four most ardent desires (in order) was

    (1) to prove the Riemann Hypothesis;

    (2) to score a century at Lords in a test match;

    (3) to prove the non-existence of God;

    (4) to assassinate Benito Mussolini.

    The list could vary slightly, but at its top was always the Riemann Hypothesis.

- On each of his regular visits to his Swedish mathematical friend Harald Bohr (younger brother of Niels and the one mentioned before on p. 56), the unswerving routine was to arrive and sit down to construct an agenda for the visit; the first point on it was always 'prove the Riemann Hypothesis'.

- On the return from one such visit, facing a stormy sea passage, he scribbled a postcard and posted it to Littlewood, which read, 'Have proved the Riemann Hypothesis'; Hardy, the atheist, reasoned that if God did exist, He would not allow him to die with the unjustified super-reputation that would have resulted in him proving this most sought after of results. Hardy arrived safely in England before the postcard arrived.

When he was asked which mathematical problem was the most important, Hilbert answered, 'The problem of the zeros of the Zeta function, not only in mathematics, but absolutely most important!'. Alternatively, one could take M. Kline's view, when he said in an interview for 'Mathematical People' in 1985:

> If I could come back after five hundred years and find that the Riemann Hypothesis or Fermat's last 'theorem' was proved, I would be disappointed, because I would be pretty sure, in view of the history of attempts to prove these conjectures, that an enormous amount of time had been spent on proving theorems that are unimportant to the life of man.

With Andrew Wiles's contribution to Fermat's Last Theorem, he must already be unhappy and there are any number of current professional and amateur mathematicians who would like to make him unhappier still!

Mathematicians do not like producing 'conditional' proofs and if they do so it shows the considerable esteem in which an unproven result is held; with this said, there are many, many results that begin: 'Assuming the truth of the Riemann Hypothesis...'. An observation by Freeman Dyson has brought about important connections with quantum theory; who knows, the greatest problem of abstract pure mathematics might be solved by a physicist—and perhaps experimentally? Certainly, fame (and now fortune) await the solver; as the advertising slogan of the British National Lottery would have it, 'It could be you', although Jonathan J. Dowling's poem (overleaf) may serve as a cautionary warning.

The Riemann Conjecture

Mein lieber Herr Riemann
All night I will dream on,
'bout how you deserve a lecture.
But of course I allude
To your famous and shrewd
Outstanding and unsolved conjecture.

Oh, I owe you my life,
My 3 kids and my wife,
For the proof of the Prime Number Theorem.
Your Zeta function trick
Made the proof really slick,
And those primes—no more do I fear 'em.

But I just stop to think,
How I've taken to drink,
And evolved this hysterical laugh-
Because still I don't know
If ζ's roots will all go
On the line real z is a half!

So I don't sleep at night,
And I'm losing my sight
In search of this darn thing's solution.
As my mind starts to go
My calculations grow
In a flood of 'complex' confusion.

I bought a computer;
Not any astuter,
It ran for nearly 10 years—no jive!
But still it *doesn't* know
If Zeta's roots *all* go
On that line real z is .5

Now I sit in my room—
I feel doomed in the gloom—
And entombed by mountains of paper.
Still, I pray that some night
My 'ol' lightbulb' will light
With the clue that could wrap up this paper.

The Greek Alphabet

A	α	alpha	a
B	β	beta	b
Γ	γ	gamma	g
Δ	δ	delta	d
E	ϵ	epsilon	e
Z	ζ	zeta	z
H	η	eta	ê
Θ	θ	theta	th
I	ι	iota	i
K	κ	kappa	k
Λ	λ	lambda	l
M	μ	mu	m
N	ν	nu	n
Ξ	ξ	xi	ks
O	o	omikron	o
Π	π	pi	p
P	ρ	rho	r
Σ	σ	sigma	s
T	τ	tau	t
Y	υ	upsilon	u
Φ	ϕ	phi	f
X	χ	chi	ch
Ψ	ψ	psi	ps
Ω	ω	omega	ô

Big Oh Notation

Introduced in 1894 by one Paul Bachmann, later embraced by number theorists in general and later still by computer scientists to measure the complexity of algorithms, this notation exposes the size of an expression while suppressing unnecessary detail.

For example, $2n^2 + 7n + 6 \to \infty$ as $n \to \infty$ but not really any more quickly that n^2 itself, since as n becomes bigger the $7n + 6$ term becomes increasingly less relevant and could be any other linear expression in n; put another way, $(2n^2 + 7n + 6)/n^2 \to 2$ as $n \to \infty$. If the 2 has no relevance, other than it being a constant, we write $2n^2 + 7n + 6 = O(n^2)$ and in general for positive functions, $g(n) = O(f(n))$ if $g(n)$ is asymptotically no bigger than a constant times $f(n)$; that is, $f(n)$ is the dominant asymptotic term of $g(n)$.

This means that $O(1)$ represents a constant and, for example, $\ln n + \ln \ln n = O(\ln n)$.

The use of the O for 'order' brings about the appropriate name of 'big oh' notation.

Taylor Expansions

The simplest functions are polynomials, since they are generally very susceptible to standard mathematical processes. If a function is not a polynomial, we can look for the best polynomial approximation to it, at least over some interval, but we must expect global difficulties; for example, the function may have a vertical asymptote or be periodic or possess any other non-polynomial behaviour. We naturally proceed by the degree of the polynomial, that is, the highest power of x.

C.1 DEGREE 1

It is intuitively clear that the best straight line that approximates a given curve at a given point is the tangent to the curve at that point (see Figure C.1). If P is the point $(a, f(a))$, the gradient of the curve at P is $f'(a)$ and the equation of the tangent is $y - f(a) = f'(a)(x - a)$ and so we have the approximation

$$f(x) \approx f(a) + (x - a)f'(a),$$

which is Taylor's first approximation.

C.2 DEGREE 2

Above we have simply used our intuition as to what the best approximating straight line would be, but we could have been more rigorous. The general straight line has two independent parameters, which together uniquely specify it: in its standard form $y = mx + c$ they are m and c. Two independent parameters means that we can impose two independent conditions on the line if we are to judge it to be the best one to achieve our approximation, and what better conditions than that the line passes through P and has the same gradient as $f(x)$ at P? In other words, the line is indeed the tangent to the curve at P. With a degree 2 approximation, we are approximating the function near P by a parabola, which in its general form $y = Ax^2 + Bx + C$ has three independent parameters A, B, C. It is perfectly natural to impose the same two conditions as before, but what of the third?

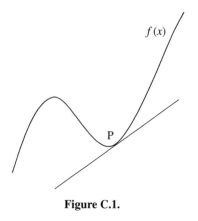

Figure C.1.

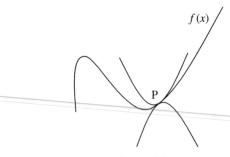

Figure C.2.

If we look at Figure C.2, we will see two parabolas being used as approximations. They both pass through P and both have the same gradient as $f(x)$ at P but we would surely prefer the upper one to the lower since it bends in the right direction. This third condition ought to distinguish between the two possibilities and since concavity is measured by the second derivative we will insist that the second derivatives of the function and of the quadratic approximation are equal at P. If we agree to write the quadratic in the more useful way of $y = A(x-a)^2 + B(x-a) + C$, we can easily evaluate the three parameters as follows:

$$\frac{dy}{dx} = 2A(x-a) + B, \qquad \frac{d^2y}{dx^2} = 2A.$$

Putting $x = a$ in the expressions for y, dy/dx and d^2y/dx^2 and imposing our conditions then gives $C = f(a)$, $B = f'(a)$, $A = \frac{1}{2}f''(a)$ and the approximation as

$$f(x) \approx f(a) + (x-a)f'(a) + \tfrac{1}{2}(x-a)^2 f''(a).$$

In general, we can continue the process by approximating by a cubic, quartic, etc., insisting that each higher derivative at P is equal to that derivative of the function at P to get

$$f(x) \approx f(a) + (x-a)f'(a) + \frac{(x-a)^2}{2!}f''(a) + \frac{(x-a)^3}{3!}f''(a) + \cdots,$$

noting that the denominators are factorials because of the repeated bringing down of the powers in the differentiation process.

C.3 EXAMPLES

$$(1+x)^\alpha \approx 1 + \alpha x + \alpha(\alpha - 1)\frac{x^2}{2!} + \alpha(\alpha - 1)(\alpha - 2)\frac{x^3}{3!} + \cdots,$$

$$e^x \approx 1 + x + \frac{x^2}{2!} + \frac{x^3}{3!} + \cdots,$$

$$\sin x \approx x - \frac{x^3}{3!} + \frac{x^5}{5!} - \cdots,$$

are easily computed, taking $a = 0$ and with this value of a the name Taylor is often replaced by the name Maclaurin. An important case where we cannot approximate taking $a = 0$ is with the function $\ln x$, since it simply is not defined there. Rather than take another value for a, it is more convenient to shift the function sideways by 1 to get $\ln(1 + x) \approx x - \frac{1}{2}x^2 + \frac{1}{3}x^3 - \frac{1}{4}x^4 + \cdots$.

C.4 CONVERGENCE

It is clear that, provided the function is infinitely differentiable, the Taylor process can be continued indefinitely (although even then there can be problems, as we will mention later) to give an infinite series rather than a polynomial and although it is designed to approximate at a point we would expect a decent approximation in some neighbourhood of the point; just how big that neighbourhood is is determined by the size of the error term involved for any degree of approximation and in particular by its asymptotic size. We will not consider this and therefore avoid Taylor's Theorem, but amazingly for a number of the important functions, the error term is asymptotically zero for all x and so the infinite series equals the function. Putting $\alpha = -1$ in the first example above results in $1/(1 + x) \approx 1 - x + x^2 - x^3 + \cdots$, which we know from the theory of geometric series is exact for $|x| < 1$ and so approximating $1/(1 + x)$ at the point $(0, 1)$ results in an exact alternative of $1 - x + x^2 - x^3 + \cdots$ in $|x| < 1$. The news is better still with, for example, e^x and $\sin x$ since the infinite series equal the function for all x, in fact, the series can be used to define such functions—and of course the series can make sense with $x \in \mathbb{R}$ replaced by $z \in \mathbb{C}$.

Complex Function Theory

D.1 COMPLEX DIFFERENTIATION

With a real-valued function of a real variable, the standard definition of the derivative is

$$f'(x) = \lim_{\delta x \to 0} \frac{f(x + \delta x) - f(x)}{\delta x}$$

given that the limit exists. It was Cauchy who provided this rigorous definition, which has great geometric appeal, as a variable chord ever more accurately approximates a given tangent; zooming in as the chord shortens forces the eye to accept that the function, the chord and the tangent are all blending into one another, making it that bit easier to believe that the final limit is indeed the gradient of the tangent to the curve at the point and, in fact, defines that tangent. It is in the direction in which $\delta x \to 0$ that the greatest subtlety lies, as the definition of derivative relies on that limit being the same no matter from which direction $\delta x \to 0$; $f(x) = |x|$ is not differentiable at the origin because of this. If we replace $x \in \mathbb{R}$ by $z \in \mathbb{C}$, we can formally write

$$f'(z) = \lim_{\delta z \to 0} \frac{f(z + \delta z) - f(z)}{\delta z}.$$

The comfortable geometric interpretation has deserted us, leaving a gap filled only by cold analysis and, as with the real case, the formula is taken as the definition of the derivative of the function at the point z. Again, if we think carefully about the $\delta z \to 0$, we now have an infinite number of directions from which to choose, rather than just the two, and if we insist that the limit does not depend on the direction, we are surely asking a great deal more than in the real case; and so it turns out. If we recall that $\mathbb{C}$ includes $\mathbb{R}$, there are three cases to consider, the first two of which we can dispose of quickly—but not the third.

D.1.1 *A real-valued function of a complex variable*

As an example, consider the simple function $f(z) = x$, where $z = x + iy$. If we approach the limit along the real axis, we get $\delta z = \delta x$ and

$$f'(z) = \lim_{\delta z \to 0} \frac{f(z + \delta z) - f(z)}{\delta z} = \lim_{\delta x \to 0} \frac{x + \delta x - x}{\delta x} = 1,$$

whereas, along the imaginary axis, $\delta z = i\delta y$ and

$$f'(z) = \lim_{\delta z \to 0} \frac{f(z + \delta z) - f(z)}{\delta z} = \lim_{i\delta y \to 0} \frac{x - x}{i\delta y} = 0.$$

So, this seemingly most simple function has no derivative. If we look at things more closely, we can identify the root of the problem: approaching the limit along real values must mean that if the limit exists it is real, whereas, approaching it along imaginary values must mean that if it exists it is imaginary, since the denominator is imaginary and the numerator is real. The only possible reconciliation is if the imaginary limit is 0, in which case, if the function is to be differentiable, the real limit must be 0 also. In summary, if such a function is differentiable, its derivative must be identically 0.

D.1.2 *A complex-valued function of a real variable*

If we write $f(x) = u(x) + iv(x)$, then

$$f'(x) = \lim_{\delta x \to 0} \frac{(u(x + \delta x) + iv(x + \delta x)) - (u(x) + iv(x))}{\delta x}$$

$$= \lim_{\delta x \to 0} \frac{u(x + \delta x) - u(x) + iv(x + \delta x) - iv(x)}{\delta x}$$

$$= \lim_{\delta x \to 0} \frac{u(x + \delta x) - u(x)}{\delta x} + i \lim_{\delta x \to 0} \frac{v(x + \delta x) - v(x)}{\delta x}$$

$$= \frac{\partial u}{\partial x} + i \frac{\partial v}{\partial x}$$

provided that the derivatives exist. The matter is therefore reduced to the two real cases.

D.1.3 *A complex-valued function of a complex variable*

We can write that if $z = x + iy$, then $f(z) = u(x, y) + iv(x, y)$. This third and final case has deep-lying consequences and lives at the heart of complex calculus—and it has its surprises. First of all, a name. Any such function, which has a meaningful derivative wherever it is defined in a region, is called *analytic* in that region (another term used is *holomorphic*) and if the region happens to be the whole of the complex plane it is said to be *entire*.

Suppose that we once again approach 0 along the real axis and then along the imaginary axis:

$$f'(z) = \lim_{\delta x \to 0} \frac{f(z + \delta x) - f(z)}{\delta x}$$

$$= \lim_{\delta x \to 0} \left\{ \frac{u(x + \delta x, y) - u(x, y)}{\delta x} + i \frac{v(x + \delta x, y) - v(x, y)}{\delta x} \right\}$$

$$= \frac{\partial u}{\partial x} + i \frac{\partial v}{\partial x}$$

and

$$f'(z) = \lim_{i\delta y \to 0} \frac{f(z + i\delta y) - f(z)}{i\delta y}$$

$$= \lim_{\delta x \to 0} \left\{ \frac{u(x, y + \delta y) - u(x, y)}{i\delta y} + i \frac{v(x, y + \delta y) - v(x, y)}{i\delta y} \right\}$$

$$= -i \frac{\partial u}{\partial y} + \frac{\partial v}{\partial y},$$

and for these to be the same we must have that

$$\frac{\partial u}{\partial x} = \frac{\partial v}{\partial y} \quad \text{and} \quad \frac{\partial u}{\partial y} = -\frac{\partial v}{\partial x}.$$

This is, of course, simply a necessary condition for the derivative to be properly defined. It turns out not quite to be sufficient, for that we need all four partial derivatives to be continuous as well. These are called the Cauchy–Riemann equations, and using them we have four equivalent ways of writing the derivative of an analytic function; in particular,

$$f'(z) = \frac{df}{dz} = \frac{\partial u}{\partial x} + i \frac{\partial v}{\partial x}.$$

It is not difficult to see that the standard rules of differentiation carry across to the complex case—linearity, product rule, quotient rule and chain rule—as do a number of reasonable general results, in particular, if $f(z) = z^n$, then $f'(z) = nz^{n-1}$ for $n \in \mathbb{R}$. More general results can carry across too, for example:

if $f'(z) = 0$ for all z, then $f(z) = c$, provided the domain is connected.

The qualification that the domain should be connected is needed even in the real case, since if

$$f(x) = \begin{cases} 0, & x < 1, \\ 1, & x > 2, \end{cases}$$

its derivative is clearly zero; the analogous complex case is

$$f(z) = \begin{cases} 0, & |z| < 1, \\ 1, & |z| > 2, \end{cases}$$

and again, clearly, $f'(z) = 0$.

Now suppose that the domain is connected.

If $f'(z) = 0$,

$$\frac{\partial u}{\partial x} + i \frac{\partial v}{\partial x} = \frac{\partial v}{\partial y} - i \frac{\partial u}{\partial y} = 0,$$

which of course means that

$$\frac{\partial u}{\partial x} = \frac{\partial v}{\partial x} = \frac{\partial v}{\partial y} = \frac{\partial u}{\partial y} = 0.$$

Since $\partial u/\partial x = 0$, $u(x, y)$ is constant along horizontal line segments; similarly, since $\partial u/\partial y = 0$, $u(x, y)$ is constant along vertical line segments. The same argument holds for $v(x, y)$. Therefore, $f(z) = u(x, y) + iv(x, y)$ is constant along each horizontal and vertical line segment in the domain. But the domain is connected and so any two points, z_1, z_2, in it can be joined by a series of horizontal and vertical line segments, which lie entirely in the domain and the function is constant along all of them, consequently $f(z_1) = f(z_2)$. Since z_1 and z_2 are arbitrary, $f(z)$ must be constant in the whole domain.

As a second reasonable general result we have that if $|f(z)| = c$, then $f(z) = c$. To establish this, use the definition of $| \cdot |$ to get $|f(z)| = c \Leftrightarrow u^2 + v^2 = c^2$. Partial differentiation with respect to x and then y gives

$$2u \frac{\partial u}{\partial x} + 2v \frac{\partial v}{\partial x} = 0 \quad \text{and} \quad 2u \frac{\partial u}{\partial y} + 2v \frac{\partial v}{\partial y} = 0.$$

Cancelling the 2 and using the Cauchy–Riemann equations gives

$$u \frac{\partial u}{\partial x} - v \frac{\partial v}{\partial y} = 0 \quad \text{and} \quad u \frac{\partial u}{\partial y} + v \frac{\partial v}{\partial x} = 0.$$

Treat these as two equations in two unknowns to get

$$(u^2 + v^2) \frac{\partial u}{\partial x} = c^2 \frac{\partial u}{\partial x} = 0,$$

so either $c = 0$ and $f(z) = 0$ identically or $\partial u/\partial x = 0$. Similarly,

$$\frac{\partial u}{\partial y} = \frac{\partial v}{\partial x} = \frac{\partial v}{\partial y} = 0.$$

Therefore, $f'(z) = 0$ and from above $f(z) = c$. Actually, the result holds if we have Re $f(z) = c$ or Im $f(z) = c$.

It is hardly surprising that the function $f(z) = z$ is differentiable—we have $u = x$ and $v = y$, making

$$\frac{\partial u}{\partial x} = \frac{\partial v}{\partial y} = 1 \quad \text{and} \quad \frac{\partial u}{\partial y} = -\frac{\partial v}{\partial x} = 0$$

—but hardly believable that $f(z) = \bar{z}$ is not (here, $u = x$ and $v = -y$, which cause the first Cauchy–Riemann equation to fail): intuition has no place in the study of the behaviour of complex functions!

If we use the ideas of Taylor expansions to extend the standard elementary functions we can formally give meaning to

$$\sin z = z - \frac{z^3}{3!} + \frac{z^5}{5!} - \cdots,$$

$$\cos z = 1 - \frac{z^2}{2!} + \frac{z^4}{4!} - \cdots,$$

$$e^z = 1 + z + \frac{z^2}{2!} + \frac{z^3}{3!} + \cdots,$$

and others like them, all of which can be shown to converge for all $z \in \mathbb{C}$. Notice that term-by-term differentiation yields the expected results

$$\frac{d}{dz}\sin z = \cos z, \qquad \frac{d}{dz}\cos z = -\sin z, \qquad \frac{d}{dz}e^z = e^z.$$

Furthermore, we have that

$$e^{iz} = \cos z + i\sin z, \qquad \sin z = \frac{e^{iz} - e^{-iz}}{2i}, \qquad \cos z = \frac{e^{iz} + e^{-iz}}{2},$$

$$\sinh z = \frac{e^z - e^{-z}}{2} = -i\sin iz, \qquad \cosh z = \frac{e^z + e^{-z}}{2} = \cos iz, \qquad \text{etc.}$$

All of these (and many more such expressions) are no more than their real counterparts with z replacing x, which begins to bring about a cosy familiarity—soon broken by the equation $\cos z = 2$ having solutions. That it does must mean that

$$\frac{e^{iz} + e^{-iz}}{2} = 2, \qquad e^{iz} + e^{-iz} = 4, \qquad e^{2iz} + 1 = 4e^{iz},$$

$$e^{iz} = \frac{4 \pm \sqrt{16 - 4}}{2} = \frac{4 \pm \sqrt{12}}{2} = 2 \pm \sqrt{3}.$$

If we allow the usual taking of logs, we are forced to write $iz = \ln(1 \pm \sqrt{3})$ and $z = -i\ln(1 \pm \sqrt{3})$, which gives a solution of $z = -i\ln(1 + \sqrt{3})$, uncomfortable because $\cos z = 2$ has a solution at all, and less comfortable still because $z = -i\ln(1 - \sqrt{3})$ is another and we recall the 'fact' that we cannot have the log of a negative number. This 'sophistry' will be explained later.

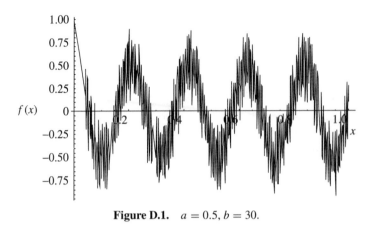

Figure D.1. $a = 0.5, b = 30$.

D.2 WEIERSTRASS FUNCTION

With our current knowledge of fractals, the idea of a real function existing which is everywhere continuous but nowhere differentiable is not novel but back in 1861 none was known, although their existence was suspected and in particular by Riemann, who suggested the idea to some of his students and even provided a candidate—but no proof. It took until 1872 when Weierstrass provided his own function to do the job—one of those events that helped to force more rigour into mathematics. In fact, he proved that for b an odd integer greater than 1 and for $0 < a < 1$, then if $ab > 1 + \frac{3}{2}\pi$, the function $f(x) = \sum_{r=1}^{\infty} a^r \cos(b^r x)$ is indeed everywhere continuous but nowhere differentiable; Hardy later extended the result to $ab \geqslant 1$ (see Figure D.1).

In the complex case we do not have to look nearly so hard to find such a monster, as the simple modulus function will do the job. We have seen that in the real case, the function causes a difficulty in that it is not differentiable at the origin, although it is obviously everywhere continuous; in the complex case, matters are much worse: $f(z) = |z|$ is a continuous, real-valued function of a complex variable and we have seen that if its derivative exists, it must be 0, which seems suspicious. In fact, its derivative exists nowhere and we can prove that using the Cauchy–Riemann equations since $u(x, y) = \sqrt{x^2 + y^2}$ and $v(x, y) = 0$, therefore

$$\frac{\partial u}{\partial x} = \frac{2x}{\sqrt{x^2 + y^2}}, \qquad \frac{\partial u}{\partial y} = \frac{2y}{\sqrt{x^2 + y^2}}, \qquad \frac{\partial v}{\partial x} = \frac{\partial v}{\partial y} = 0$$

and, as long as not both x and y are 0, the Cauchy–Riemann equations are clearly not satisfied; the case $x = y = 0$ gives the indeterminate form $0/0$ and we have to go back to first principles to get

$$\frac{\partial u}{\partial x}\bigg|_{(0,0)} = \lim_{h \to 0} \frac{u(h, 0) - u(0, 0)}{h}$$

$$= \lim_{h \to 0} \frac{\sqrt{h^2}}{h} = \lim_{h \to 0} \frac{|h|}{h} = \begin{cases} 1, & h \to 0^+, \\ -1, & h \to 0^-, \end{cases}$$

just as with $f(x) = |x|$. The same argument shows that

$$\frac{\partial u}{\partial y}\bigg|_{(0,0)}$$

does not exist.

The function $f(z) = |z|$ is clearly complicated, but it is not as bad as its companion $f(z) = \arg z$, which is not even properly defined, as it is only determined up to integer multiples of 2π; when it is restricted to $[-\pi, \pi]$, it is usually written with a capital 'A', and then $f(z) = \operatorname{Arg} z = \tan^{-1}(y/x)$. Once again, it is a real-valued function of a complex variable and so we know that if its derivative exists anywhere, it must be zero and if we apply the Cauchy–Riemann equations once more we get

$$u(x, y) = \tan^{-1}\left(\frac{y}{x}\right), \qquad v(x, y) = 0,$$

$$\frac{\partial u}{\partial x} = \frac{-y}{x^2 + y^2}, \qquad \frac{\partial u}{\partial y} = \frac{x}{x^2 + y^2}, \qquad \frac{\partial v}{\partial x} = \frac{\partial v}{\partial y} = 0,$$

and once again we have that, as long as not both x and y are 0, the Cauchy–Riemann equations are not satisfied; since the function is not defined at $x = y = 0$, it is nowhere differentiable.

D.3 COMPLEX LOGARITHMS

We can define the complex logarithm by its formal power series

$$\ln(1 + z) = z - \tfrac{1}{2}z^2 + \tfrac{1}{3}z^3 - \cdots, \qquad |z| \leqslant 1,$$

and, just as in the real case,

$$\ln\left(\frac{1 + z}{1 - z}\right) = 2(z + \tfrac{1}{3}z^3 + \tfrac{1}{5}z^5 + \cdots), \qquad |z| < 1,$$

if we want to reach complex numbers outside the unit circle, but that disguises the important subtlety brought about by the ambiguous nature of the argument of a complex number. If we take the definition of the logarithm as the inverse of the exponential function, the matter is much more clear. So, write $w = \ln z$

if $z = e^w$. If $w = u + iv$ and $z = r(\cos\theta + i\sin\theta)$, we have that $z = e^w = e^{u+iv} = e^u e^{iv} = e^u(\cos v + i\sin v) = r(\cos\theta + i\sin\theta)$, giving two expressions for z and, in particular, $|z|$ to give $e^u = r$ and hence $u = \ln r$, a genuine real logarithm. Also, $\cos v + i\sin v = \cos\theta + i\sin\theta$, which means that $\cos v = \cos\theta$ and $\sin v = \sin\theta$ must both be satisfied and so $v = \theta + 2n\pi$, where $n \in \mathbb{Z}$. All of this means that $\ln z = \ln r + i(\theta + 2n\pi)$ is a multivalued function. Restricting to the principal arg function Arg makes $n = 0$ and the principal logarithm function is written $\ln z = \ln r + i\theta$ for $-\pi < \theta \leqslant \pi$, or $\ln z = \ln|z| + i\,\mathrm{Arg}\,z$. In the series above, the lowercase 'l' should be replaced by its capital. The earlier solution $z = -i\ln(1 - \sqrt{3})$ of $\cos z = 2$ is then $z = i(\ln 2 + i(-\tfrac{1}{3}\pi)) = \tfrac{1}{3}\pi + i\ln 2$.

Now we can differentiate $\ln z$ in the usual way:

$$\ln z = \ln\sqrt{x^2 + y^2} + i\tan^{-1}\left(\frac{y}{x}\right)$$

so $u(x, y) = \tfrac{1}{2}\ln(x^2 + y^2)$, $v(x, y) = \tan^{-1}(y/x)$ and

$$\frac{\partial u}{\partial x} = \frac{x}{x^2 + y^2}, \qquad \frac{\partial u}{\partial y} = \frac{y}{x^2 + y^2},$$

$$\frac{\partial v}{\partial x} = \frac{-y}{x^2 + y^2}, \qquad \frac{\partial v}{\partial y} = \frac{x}{x^2 + y^2}.$$

The Cauchy–Riemann equations are satisfied and

$$\frac{d}{dz}\ln z = \frac{x}{x^2 + y^2} - i\frac{y}{x^2 + y^2} = \frac{1}{z},$$

as we might have hoped.

The mixture of surprise and familiarity is an inevitable part of the demanding definition of complex differentiability and it would be reasonable to think that, with its lesser demand of continuity, complex integration would be more predictable in its behaviour—but once again intuition fails us.

D.4 COMPLEX INTEGRATION

D.4.1 *The definite integral*

Before we can properly discuss complex integration we need to understand the topological idea of a region being 'simply connected', which really means that it has no holes. Put more mathematically, we will say that a region is simply connected if any closed curve drawn in it can be continuously deformed to any other closed curve in it, without leaving the region; we have already met this idea on p. 227. Geometrically, see Figure D.2.

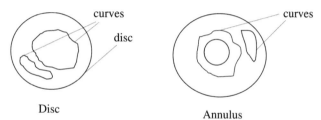

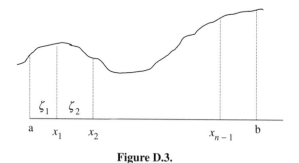

Figure D.2.

Figure D.3.

Clearly, any two closed curves drawn inside the disc can be continuously deformed into one another without leaving the disc, but the two drawn in the annulus cannot be. Another way of saying the same thing is that in a simply connected region, while staying inside the region, any closed curve can be shrunk to a point. Two other definitions will also be useful to us: a curve is said to be *simple* if it does not touch itself or self-intersect and it is said to be *smooth* if it has a well-defined tangent at every point. Now to the theory.

As with differentiation, the definition of the complex definite integral relies heavily on its real counterpart and so it is sensible if we look at that first. Suppose that $f(x)$ is a continuous real-valued function of a real variable, defined for $a \leqslant x \leqslant b$, divide the interval $[a, b]$ by introducing the points $a = x_0, x_1, x_2, \ldots, x_n = b$.

Then we define

$$S_n = \sum_{r=1}^{n} f(\xi_r)(x_r - x_{r-1}) = \sum_{r=1}^{n} f(\xi_r)\delta x_r,$$

where ξ_r is any point in the interval $[x_{r-1}, x_r]$, as the sum of areas of rectangles approximating the area under the curve; $S_n \to \int_a^b f(x)\,dx$ in the limit as $n \to \infty$.

Now suppose that we have a smooth curve C defined in the complex plane, which starts at the point a and continues to the point b, and is divided by points

233

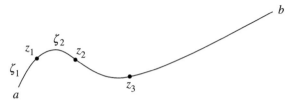

Figure D.4.

$a = z_0, z_1, z_2, \ldots, z_n = b$. Introduce the interior points ζ_r to get Figure D.4 and

$$S_n = \sum_{r=1}^{n} f(\xi_r)(z_r - z_{r-1}) = \sum_{r=1}^{n} f(\xi_r)\delta z_r.$$

Now take the limit as $n \to \infty$ to get

$$S_n \to \oint_C f(z)\,dz.$$

The geometric interpretation of areas of rectangles ever better approximating the area under a curve is lost, but we have a formal and natural extension of the idea.

If we represent C in the parametrized form $z(t) = x(t) + iy(t)$, where $z(\alpha) = a$ and $z(\beta) = b$ and rewrite the expression in a slightly different way, we get

$$\sum f(z(t))(z(t + \delta t) - z(t)) = \sum f(z(t))\frac{z(t + \delta t) - z(t)}{\delta t}\delta t$$
$$\xrightarrow[\delta t \to 0]{} \int_\alpha^\beta f(z(t))\frac{dz(t)}{dt}\,dt,$$

which makes it clear why the curve needs to be smooth. In short, we have

$$\oint_C f(z)\,dz = \int_\alpha^\beta f(z)\frac{dz}{dt}\,dt.$$

The standard rules of linearity are inherited from the $\sum$ to give

$$\oint_C f_1(z) + f_2(z)\,dz = \oint_C f_1(z)\,dz + \oint_C f_2(z)\,dz$$

and

$$\oint_C \zeta f(z)\,dz = \zeta \oint_C f(z)\,dz \quad \text{for } \zeta \in \mathbb{C};$$

for the same reason, if C is made up from two smooth curves, C_1 and C_2,

$$\oint_C f(z)\,dz = \oint_{C_1} f(z)\,dz + \oint_{C_2} f(z)\,dz$$

and further that

$$\oint_C f(z)\,dz = -\oint_C f(z)\,dz,$$

where the arrows indicate the direction in which C is traversed.

D.5 A Useful Inequality

Suppose that $|f(z)| \leqslant M$ for all $z \in C$ and that C has length L, then

$$|S_n| = \left|\sum_{r=1}^{n} f(\xi_r)\delta z_r\right| \leqslant \sum_{r=1}^{n} |f(\xi_r)||\delta z_r| \leqslant M \sum_{r=1}^{n} |\delta z_r|.$$

Since $|\delta z_r|$ is the length of the chord joining z_r and z_{r-1}, as $n \to \infty$,

$$\sum_{r=1}^{n} |\delta z_r| \to L,$$

by definition of the length of a curve, and we have the result that

$$\left|\oint_C f(z)\,dz\right| \leqslant ML.$$

D.6 The Indefinite Integral

With real-valued functions of a real variable, integration is, of course, the process of finding the (signed) area under the graph of a function, but is also the process of anti-differentiation and the two are linked by the Fundamental Theorem of Analysis, which states that

$$\int_a^b f(x)\,dx = [F(x)]_a^b = F(b) - F(a),$$

where $F(x)$ is defined as any function such that $dF(x)/dx = f(x)$, in which case, $F(x)$ is called the 'indefinite integral' of $f(x)$ and is written $F(x) = \int f(x)\,dx$.

With this result in place, we know that finding the area under a curve becomes a matter of anti-differentiation; for example,

$$\int_0^1 x\,dx = \left[\frac{x^2}{2}\right]_0^1 = \frac{1^2}{2} - \frac{0^2}{2} = \frac{1}{2}.$$

It would be nice if we could do the same in the complex case to get, for example,

$$\int_0^{1+i} z\,dz = \left[\frac{z^2}{2}\right]_0^{1+i} = \frac{(1+i)^2}{2} - \frac{0^2}{2} = i.$$

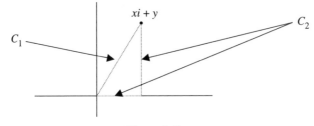

Figure D.5.

In the real case, there is never any choice about how the upper limit is approached from the lower, the crucial point here is that the result would have no regard for the infinite number of paths that could be taken to get from 0 to $1 + i$. If we have a Fundamental Theorem of Calculus in the complex case, the path has to be irrelevant, which seems an overly optimistic hope—but consider the trivial function $f(z) = 1$ integrated over any path connecting the points a and b:

$$\oint_C 1\,dz = \lim_{n\to\infty} ((z_1 - a)1 + (z_2 - z_1)1 + (z_3 - z_2)1 + \cdots + (z_n - z_{n-1})1)$$

$$= \lim_{n\to\infty} (z_n - z_0) = b - a.$$

The path is indeed irrelevant and we could write

$$\oint_C 1\,dz = \int_a^b 1\,dz = [z]_a^b = b - a.$$

A promising start, but things soon go wrong.

For example, if $f(z) = \mathrm{Re}(z) = x$ and we integrate from $a = 0$ to $b = x + iy$ along C_1 and C_2 as shown in Figure D.5 we get

$$C_1: \quad z(t) = xt + iyt, \quad 0 \leqslant t \leqslant 1,$$

to give $\dot{z}(t) = x + iy$ and

$$\int_{C_1} \mathrm{Re}(z)\,dz = \int_0^1 xt(x + iy)\,dt = \tfrac{1}{2}x(x + iy) = \tfrac{1}{2}x^2 + i\tfrac{1}{2}xy;$$

$$C_2: \quad z_1(t) = t, \quad 0 \leqslant t \leqslant x, \quad z_2(t) = x + it, \quad 0 \leqslant t \leqslant y,$$

and so $\dot{z}_1(t) = 1$ and $\dot{z}_2(t) = i$ to give

$$\int_{C_2} \mathrm{Re}(z)\,dz = \int_0^x t.1\,dt + \int_0^y xi\,dt = \tfrac{1}{2}x^2 + ixy,$$

which is hardly the same answer!

Now consider, $f(z) = 1/z$, which we can consider in two separate ways: it is defined and analytic in any annular region centred on the origin, or it is not defined at the origin and therefore not analytic in any disc centred on the origin. Suppose that we allow the annular region and the disc to contain the unit circle C, defined by $|z| = 1$, then we have

$$z(t) = \cos t + i \sin t, \qquad \dot{z}(t) = -\sin t + i \cos t$$

and

$$\oint_C \frac{1}{z}\,dz = \int_0^{2\pi} \frac{1}{\cos t + i \sin t}(-\sin t + i \cos t)\,dt$$

$$= \int_0^{2\pi} \frac{i}{\cos t + i \sin t}(\cos t + i \sin t)\,dt = 2\pi i,$$

perhaps not at first surprising, but this is closed contour and if the answer simply depended on the end points, it should be zero.

D.7 THE SEMINAL RESULT

We will not prove the result, but the reconciliation is found in

---

Cauchy's Integral Theorem

If $f(z)$ is analytic inside a simply connected domain Δ, then

$$\oint_C f(z)\,dz \text{ is constant for any contour } C \text{ lying inside } \Delta.$$

From this is follows that, if the contour joins a and b,

$$\oint_C f(z)\,dz = \int_a^b f(z)\,dz = F(b) - F(a),$$

where $F(z) = \int^z f(\zeta)\,d\zeta$ is the indefinite integral.

Notice also that this implies that if C is a closed contour, $\oint_C f(z)\,dz = 0$ and the example with $f(z) = 1/z$ above demonstrates that the analytic and simply connected conditions are both necessary.

From this we can see that

$$\oint_C f(z)\,dz = \oint_{C_1} f(z)\,dz$$

if C_1 is any path continuously deformed from C, with its ends fixed; this is called the Principle of Deformation of Path.

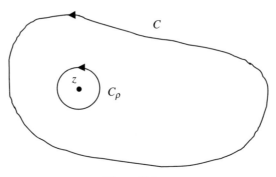

Figure D.6.

D.8 AN ASTONISHING CONSEQUENCE

Cauchy's Integral Formula

If $f(z)$ is analytic inside a simply connected domain Δ, then for any point $z \in \Delta$ and any simple closed contour $C \in \Delta$,

$$f(z) = \frac{1}{2\pi i} \oint_C \frac{f(\zeta)}{\zeta - z} \, d\zeta.$$

This means that at every point of the domain the function is determined by its values on any simple, closed contour in the domain and enclosing the point; a result that has a most peculiar feel to it. To prove it, draw a circle C_ρ of radius ρ around z, then by the deformation of path principle (see Figure D.6),

$$\oint_C \frac{f(\zeta)}{\zeta - z} \, d\zeta = \oint_{C_\rho} \frac{f(\zeta)}{\zeta - z} \, d\zeta.$$

We now evaluate the right-hand side of the above expression:

$$\oint_{C_\rho} \frac{f(\zeta)}{\zeta - z} \, d\zeta = \oint_{C_\rho} \frac{f(\zeta) - f(z)}{\zeta - z} \, d\zeta + \oint_{C_\rho} \frac{f(z)}{\zeta - z} \, d\zeta$$

$$= \oint_{C_\rho} \frac{f(\zeta) - f(z)}{\zeta - z} \, d\zeta + f(z) \oint_{C_\rho} \frac{1}{\zeta - z} \, d\zeta.$$

By a simple translation, we know from before that

$$\oint_{C_\rho} \frac{1}{\zeta - z} \, d\zeta = 2\pi i.$$

Now note that $(f(\zeta) - f(z))/(\zeta - z)$ is bounded for all $\zeta \neq z$ inside and on C and that

$$\lim_{\zeta \to z} \frac{f(\zeta) - f(z)}{\zeta - z} = f'(z),$$

which is finite since $f(z)$ is analytic. Therefore,

$$\left| \frac{f(\zeta) - f(z)}{\zeta - z} \right| < M \quad \text{for all } \zeta \text{ inside and on } C.$$

Consequently,

$$\left| \oint_{C_\rho} \frac{f(\zeta) - f(z)}{\zeta - z} \, d\zeta \right| \leqslant \oint_{C_\rho} \left| \frac{f(\zeta) - f(z)}{\zeta - z} \right| d\zeta$$

$$< \oint_{C_\rho} M \, d\zeta = 2\pi\rho M \xrightarrow[\rho \to 0]{} 0.$$

So

$$\oint_C \frac{f(\zeta)}{\zeta - z} \, d\zeta = 2\pi i f(z) \quad \text{and} \quad f(z) = \frac{1}{2\pi i} \oint_C \frac{f(\zeta)}{\zeta - z} \, d\zeta,$$

as required.

In a sense, this means that any analytic function $f(z)$ can be expressed in terms of a simple reciprocal function $1/(\zeta - z)$, which has far-reaching implications. For example, *an analytic function has derivatives of all orders.*

Again, this contrasts starkly with the real case, in which the differentiability of a function can easily come to an end; for example, $f(x) = x|x|$ differentiates to $f'(x) = 2|x|$.

The proof is trivial, if we allow repeated differentiation under the integral sign (which is not hard to justify). Pick any closed contour C in which $f(z)$ is analytic and write

$$f(z) = \frac{1}{2\pi i} \oint_C \frac{f(\zeta)}{\zeta - z} \, d\zeta$$

to give

$$f'(z) = \frac{1}{2\pi i} \oint_C \frac{f(\zeta)}{(\zeta - z)^2} \, d\zeta,$$

$$f''(z) = \frac{1}{2\pi i} \oint_C \frac{f(\zeta)}{(\zeta - z)^3} \, d\zeta, \quad \text{etc.}$$

With this result, we can develop a part of the theory of expansions of analytic functions.

D.9 TAYLOR EXPANSIONS — AND AN IMPORTANT CONSEQUENCE

If we define the infinitely differentiable real function,

$$f(x) = \begin{cases} e^{-1/x^2}, & x \neq 0, \\ 0, & x = 0, \end{cases}$$

239

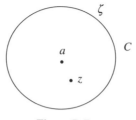

Figure D.7.

and evaluate $f(0)$, $f'(0)$, $f''(0)$, ... by taking the limit each term will be 0, with the exponential components dominating the powers of x. As a result, even though the function is infinitely differentiable, it is impossible to represent it as a Taylor series centred on $x = 0$ but once again in the complex case, the severe restriction of analyticity brings with it a stronger result. In this short section we establish the Taylor expansion and several times build on results to arrive at a result of huge significance.

Assume that $f(z)$ is analytic inside and on a circle C, centred at $z = a$; let z lie inside the circle and ζ on it (see Figure D.7). Since $\zeta - z = (\zeta - a) - (z - a)$,

$$\frac{1}{\zeta - z} = \frac{1}{(\zeta - a) - (z - a)}$$

$$= \frac{1}{\zeta - a} \frac{1}{1 - (z - a)/(\zeta - a)}$$

$$= \frac{1}{\zeta - a}\left(1 - \frac{z - a}{\zeta - a}\right)^{-1}.$$

Clearly, $|z - a| < |\zeta - a|$ and so $|(z - a)/(\zeta - a)| < 1$ and the infinite binomial expansion is valid to give

$$\frac{1}{\zeta - z} = \frac{1}{\zeta - a}\left(1 + \frac{z - a}{\zeta - a} + \left(\frac{z - a}{\zeta - a}\right)^2 \right.$$

$$\left. + \left(\frac{z - a}{\zeta - a}\right)^3 + \cdots + \left(\frac{z - a}{\zeta - a}\right)^n + \cdots\right)$$

and

$$f(z) = \frac{1}{2\pi i}\oint_C \frac{f(\zeta)}{\zeta - a}\left\{1 + \frac{z - a}{\zeta - a} + \left(\frac{z - a}{\zeta - a}\right)^2 \right.$$

$$\left. + \left(\frac{z - a}{\zeta - a}\right)^3 + \cdots + \left(\frac{z - a}{\zeta - a}\right)^n + \cdots\right\}d\zeta,$$

$$f(z) = \frac{1}{2\pi i} \oint_C \frac{f(\zeta)}{\zeta - a} \, d\zeta + \frac{(z-a)}{2\pi i} \oint_C \frac{f(\zeta)}{(\zeta - a)^2} \, d\zeta$$

$$+ \frac{(z-a)^2}{2\pi i} \oint_C \frac{f(\zeta)}{(\zeta - a)^3} \, d\zeta + \cdots + \frac{(z-a)^n}{2\pi i} \oint_C \frac{f(\zeta)}{(\zeta - a)^{n+1}} \, d\zeta$$

$$+ \cdots$$

$$= f(a) + (z-a)f'(a) + \frac{(z-a)^2}{2!} f''(a)$$

$$+ \frac{(z-a)^3}{3!} f'''(a) + \cdots + \frac{(z-a)^n}{n!} f^{(n)}(a) \cdots$$

$$= \sum_{r=0}^{\infty} A_r (z-a)^r$$

where

$$A_r = \frac{1}{2\pi i} \oint_C \frac{f(\zeta)}{(\zeta - a)^{r+1}} \, d\zeta.$$

Again, the term-by-term integration can be easily justified and we have a guaranteed (and unique) convergent Taylor expansion of the function in the disc. The formal series definitions of some of the standard functions we mentioned earlier can be made rigorous in this way.

Combine this with the 'ML' result on p. 235 and we see that the coefficients A_r satisfy the inequality

$$|A_r| = \left| \frac{1}{2\pi i} \oint_C \frac{f(\zeta)}{(\zeta - a)^{r+1}} \, d\zeta \right| \leqslant \frac{1}{2\pi} \frac{M}{\rho^{r+1}} 2\pi\rho = \frac{M}{\rho^r},$$

where $|f(\zeta)| \leqslant M$ on C, which has radius ρ.

The very reasonable earlier result that $|f(z)| = c \Rightarrow f(z) = k$ extends to a very surprising one that again simply is not true in the real case; the function $f(x) = 1/(1 + x^2)$, for example, is infinitely differentiable and bounded by 1, but it certainly isn't constant, whereas in the complex case we have that a bounded entire function is constant (this is Liouville's Theorem), which is now easy to prove.

Since the function is entire, we can expand it as a Taylor series about 0 to get $f(z) = \sum_{r=0}^{\infty} A_r z^r$. Since $f(z)$ is bounded in $\mathbb{C}$, $|f(z)| \leqslant M$ on any disc centred on 0 and of radius ρ and so $|A_r| \leqslant M/\rho^r$ for $r \geqslant 1$. Since we may take ρ arbitrarily small, $|A_r| = A_r = 0$ for $r \geqslant 1$ and $f(z) = A_0$, a constant.

And having proved that it is but a small step to one of the cornerstones of the whole of mathematics:

The Fundamental Theorem of Algebra

Any polynomial with coefficients in $\mathbb{C}$ has a root in $\mathbb{C}$.

Write the polynomial as $P(z) = a_0 + a_1 z + a_2 z^2 + \cdots + a_n z^n$. If $P(z)$ has no roots, $f(z) = 1/P(z)$ is an entire function and for $|z|$ sufficiently big (say, $|z| > R$), $|P(z)| > 1$ and so $|f(z)| < 1$. In the disc $|z| \leq R$, $|f(z)|$ is clearly continuous and it is a standard topological result that it is therefore bounded, consequently, $f(z)$ is bounded in the whole of $\mathbb{C}$ and using Liouville's Theorem it is constant. Having established one root in $\mathbb{C}$ we can reduce the degree of the polynomial by 1 by factorization and repeat the process to get the result that any polynomial of degree n with coefficients in $\mathbb{C}$ has precisely n roots in $\mathbb{C}$.

Seeing this result so neatly and easily proved belies the difficulty that was encountered initially to establish it, a task not made the easier by the mathematicians who attempted it having the deepest suspicions about complex numbers. In one of the most significant PhD theses ever, Gauss gave a first satisfactory proof of the result in 1799, albeit for real coefficients, following incomplete attempts by Descartes, Euler, d'Alembert and Lagrange; in fact, over the course of his lifetime he produced four different proofs, the last one finally dealing with the case of complex coefficients.

D.10 LAURENT EXPANSIONS—AND ANOTHER IMPORTANT CONSEQUENCE

Taylor expansion crucially needed analyticity in a simply connected region, but suppose that the region was not simply connected or that the function was not everywhere analytic? For more than 20 years from 1821 Cauchy had developed complex function theory virtually alone, until at last some of his fellow countryman began to mine the many rich ideas that he had exposed and in 1843 Pierre-Alphonse Laurent (1813–1854) answered this question by extending the idea of Taylor series to what has become appropriately known as Laurent series (Weierstrass had known about this in 1841 but had failed to publish his findings). As with the function $f(z) = 1/z$, the result can either be looked at as a series expansion of an analytic function in a region comprising a disc with a hole in it or of a function defined on a disc but having an isolated singularities—in which case we can 'cut it out' by surrounding it with a removable circle (see Figure D.8).

Suppose that the singular point z_0 is surrounded by an inner circle C_ρ and that we perform a radial cut from C to C_ρ, thereby constructing a contour which takes us all around C, radially inwards to C_ρ, all around that (in the opposite direction) and back along the radial line and then along to the start on C. This results in a simply connected region in which $f(z)$ is analytic and so we can apply Cauchy's integral formula to get

$$f(z) = \frac{1}{2\pi i} \oint_C \frac{f(\zeta)}{\zeta - z}\, d\zeta - \frac{1}{2\pi i} \oint_{C_\rho} \frac{f(\zeta)}{\zeta - z}\, d\zeta,$$

the two equal and opposite contributions from the radial parts having cancelled out.

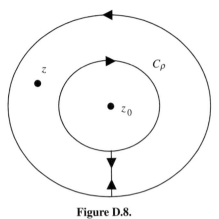

Figure D.8.

For $\zeta \in C$, $|z - z_0| < |\zeta - z_0|$ and the same reasoning as was given for the Taylor expansion gives

$$\frac{1}{2\pi i} \oint_C \frac{f(\zeta)}{\zeta - z} \, d\zeta = \sum_{r=0}^{\infty} a_r (z - z_0)^r,$$

where

$$a_r = \frac{1}{2\pi i} \oint_C \frac{f(\zeta)}{(\zeta - z_0)^{r+1}} \, d\zeta.$$

The problem with the second integral is that $\zeta \in C_\rho$ and so $|z - z_0| > |\zeta - z_0|$ and the geometric series that was developed will diverge—so we turn things upside down, since we can also write

$$\zeta - z = (\zeta - z_0) - (z - z_0),$$

and therefore

$$\begin{aligned}
\frac{1}{\zeta - z} &= \frac{1}{(\zeta - z_0) - (z - z_0)} \\
&= \frac{1}{z - z_0} \frac{-1}{1 - (\zeta - z_0)/(z - z_0)} \\
&= \frac{-1}{z - z_0} \left(1 - \frac{\zeta - z_0}{z - z_0} \right)^{-1} \\
&= \frac{-1}{z - z_0} \left(1 + \frac{\zeta - z_0}{z - z_0} + \left(\frac{\zeta - z_0}{z - z_0} \right)^2 \right. \\
&\qquad \left. + \left(\frac{\zeta - z_0}{z - z_0} \right)^3 + \cdots + \left(\frac{\zeta - z_0}{z - z_0} \right)^n + \cdots \right),
\end{aligned}$$

243

$$\frac{-1}{2\pi i}\oint_{C_\rho}\frac{f(\zeta)}{\zeta-z}\,\mathrm{d}\zeta$$

$$=\frac{1}{2\pi i}\oint_{C_\rho}\frac{f(\zeta)}{z-z_0}\left(1+\frac{\zeta-z_0}{z-z_0}+\left(\frac{\zeta-z_0}{z-z_0}\right)^2\right.$$

$$\left.+\left(\frac{\zeta-z_0}{z-z_0}\right)^3+\cdots+\left(\frac{\zeta-z_0}{z-z_0}\right)^n+\cdots\right)\mathrm{d}\zeta$$

$$=\frac{1}{2\pi i}\left\{\frac{1}{z-z_0}\oint_{C_\rho}f(\zeta)\,\mathrm{d}\zeta+\frac{1}{(z-z_0)^2}\oint_{C_\rho}(\zeta-z_0)f(\zeta)\,\mathrm{d}\zeta+\cdots\right.$$

$$\left.+\frac{1}{(z-z_0)^n}\oint_{C_\rho}(\zeta-z_0)^{n-1}f(\zeta)\,\mathrm{d}\zeta+\cdots\right\}$$

$$=\sum_{r=1}^{\infty}\frac{b_r}{(z-z_0)^r},$$

where

$$b_r=\frac{1}{2\pi i}\oint_{C_\rho}(\zeta-z_0)^{r-1}f(\zeta)\,\mathrm{d}\zeta.$$

All of this makes

$$f(z)=\sum_{r=0}^{\infty}a_r(z-z_0)^r+\sum_{r=1}^{\infty}\frac{b_r}{(z-z_0)^r}$$

the promised Laurent series of the function.

It is important to note that, just as the Taylor expansion for a given function is unique in its disc of convergence, so the Laurent expansion is unique in its annulus of convergence, although it can vary over concentric annuli. There are any number of examples of this phenomenon, for example,

$$\frac{1}{z(1+z)}=\frac{1}{z}(1-z+z^2-z^3+\cdots)$$

$$=\frac{1}{z}-1+z-z^2-\cdots,\qquad 0<|z|<1,$$

$$\frac{1}{z(1+z)}=\frac{1}{z^2(1+1/z)}=\frac{1}{z^2}\left(1-\frac{1}{z}+\frac{1}{z^2}-\frac{1}{z^3}\cdots\right)$$

$$=\frac{1}{z^2}-\frac{1}{z^3}+\frac{1}{z^4}-\cdots,\qquad 1<|z|<2,$$

where the right-hand boundary of 2 is arbitrary. Laurent series have their uses, just as Taylor series have their uses, but in pole position among them is their application to the calculation of what are known as *residues* and through that to the evaluation of real and complex definite integrals.

D.11 THE CALCULUS OF RESIDUES

Consider a function $f(z)$ defined and analytic in a domain Δ apart from a singularity at $z = z_0$ (called a pole, which explains the earlier pun). Construct a circle with z_0 as its centre, then if C is any closed contour in Δ surrounding that circle, $f(z)$ has a Laurent expansion as above in the annulus and the coefficient of the first negative power term is

$$b_1 = \frac{1}{2\pi i} \oint_C f(\zeta)\,d\zeta$$

and so

$$\oint_C f(\zeta)\,d\zeta = 2\pi i b_1;$$

consequently, if we can find the value of b_1, we can evaluate the integral. It is customary to call b_1 the *residue* of $f(z)$ at $z = z_0$ and write it as $b_1 = \operatorname{Res}_{z=z_0} f(z)$ and so we have that

$$\oint_C f(\zeta)\,d\zeta = 2\pi i \operatorname*{Res}_{z=z_0} f(z).$$

By constructing circles around each singularity individually, the idea easily generalizes to n singularities to give

$$\oint_C f(\zeta)\,d\zeta = 2\pi i \sum_{r=1}^{n} \operatorname*{Res}_{z=z_r} f(z),$$

which is known as Cauchy's Residue Theorem. Now all we need are methods to calculate the residues, of which there are many, and we will be able to evaluate the integral.

We will assume that we have a simple pole, that is, one for which the Laurent expansion has just one negative-power term and look at two related methods.

1. The Laurent series is

$$f(z) = \frac{b_1}{z - z_0} + a_0 + a_1(z - z_0) + a_2(z - z_0)^2 + \cdots .$$

Multiplying both sides by $(z - z_0)$ gives

$$(z - z_0)f(z) = b_1 + (z - z_0)\{a_0 + a_1(z - z_0) + a_2(z - z_0)^2 + \cdots\}.$$

And so

$$\operatorname*{Res}_{z=z_0} f(z) = b_1 = \lim_{z \to z_0} (z - z_0)f(z).$$

As an example, if $f(z) = \sin z/(z^2 + 1)$,

$$\operatorname*{Res}_{z=i} f(z) = \lim_{z \to i}(z - i)\frac{\sin z}{z^2 + 1} = \frac{\sin i}{2i} = \tfrac{1}{2}\sinh 1$$

and

$$\operatorname*{Res}_{z=-i} f(z) = \lim_{z \to -i} (z+i) \frac{\sin z}{z^2 + 1} = \frac{\sin(-i)}{-2i} = \tfrac{1}{2}\sinh 1.$$

If we integrate around a contour that does not contain $z = \pm i$, the function is analytic and the integral must therefore be 0, but if we integrate around $C = \{z : |z| = 2\}$,

$$\int_C \frac{\sin z}{z^2 + 1}\, dz = 2\pi i (\tfrac{1}{2}\sinh 1 + \tfrac{1}{2}\sinh 1) = (2\pi \sinh 1)i.$$

2. In the first example, the denominator of the fraction was easily factorized; suppose now that we have a rational function of z in which this is not the case. Write $f(z) = p(z)/q(z)$, where $p(z)$ and $q(z)$ are analytic. Suppose that $f(z)$ has a simple pole at $z = z_0$, so that $p(z_0) \neq 0$ and z_0 is a simple zero of $q(z)$. Expand $q(z)$ as a Taylor series about $z = z_0$ to give

$$q(z) = q(z_0) + (z - z_0)q'(z_0) + \frac{(z - z_0)^2}{2!}q''(z_0) + \cdots$$

$$= (z - z_0)q'(z_0) + \frac{(z - z_0)^2}{2!}q''(z_0) + \cdots$$

$$= (z - z_0)\left\{ q'(z_0) + \frac{(z - z_0)}{2!}q''(z_0) + \cdots \right\}.$$

So,

$$\operatorname*{Res}_{z=z_0} f(z) = b_1 = \lim_{z \to z_0} (z - z_0) f(z)$$

$$= \lim_{z \to z_0} (z - z_0) \frac{p(z)}{q(z)}$$

$$= \lim_{z \to z_0} \cancel{(z - z_0)} \frac{p(z)}{\cancel{(z - z_0)}\{q'(z_0) + ((z - z_0)/2!)q''(z_0) + \cdots\}}$$

$$= \frac{p(z_0)}{q'(z_0)}.$$

For example, if $f(z) = (z^2 + 1)/\sin z$,

$$\operatorname*{Res}_{z=0} f(z) = \frac{0^2 + 1}{\cos 0} = 1$$

and, more generally,

$$\operatorname*{Res}_{z=k\pi} f(z) = \frac{(k\pi)^2 + 1}{\cos k\pi} = \begin{cases} (k\pi)^2 + 1, & k \text{ even}, \\ -((k\pi)^2 + 1), & k \text{ odd}. \end{cases}$$

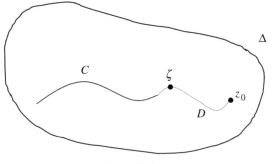

Figure D.9.

D.12 ANALYTIC CONTINUATION

Recall that the result is as follows.

> If, in some complex domain Δ, two analytic functions are defined and are equal at all points on a curve C lying inside Δ, they are equal throughout Δ.

We can now prove it as follows.

Let the two analytic functions be $f_1(z)$ and $f_2(z)$ defined in some region Δ of $\mathbb{C}$ and write their difference as $\varphi(z) = f_1(z) - f_2(z)$. Then $\varphi(z)$ is analytic throughout Δ and identically 0 on C. Now suppose that there is a point $z_0 \in \Delta$ at which $\varphi(z_0) \neq 0$; clearly, $z_0 \notin C$. Now extend C inside Δ by a curve D, heading towards z_0, and let ζ be the last point on D for which $\varphi(z) \neq 0$, then $\zeta \neq z_0$ and on the segment of the curve D beyond ζ, $\varphi(z) \neq 0$, by definition of ζ. If we differentiate $\varphi(z)$ at points on the curve up to ζ by taking the limit along the curve, we must have $\varphi(z) = \varphi'(z) = \varphi''(z) = \cdots = 0$ and, in particular, $\varphi(\zeta) = \varphi'(\zeta) = \varphi''(\zeta) = \cdots = 0$. Now expand $\varphi(z)$ as a Taylor series about the point $z = \zeta$, then all of the coefficients are 0 and so $\varphi(z) = 0$ in some circle centred at $z = \zeta$ and consequently on some of the curve beyond ζ, which is a contradiction and the result is established (see Figure D.9).

In general, how a given function achieves its continuation (if indeed it has one) depends on the function, and there can be any number of equivalent ways, leading to expressions that look different but must in fact be the same.

Application to the Zeta Function

E.1 ZETA ANALYTICALLY CONTINUED

In the first part of his paper, Riemann performed the analytic continuation of the Euler Zeta function

$$\zeta(x) = \sum_{r=1}^{\infty} \frac{1}{r^x},$$

which we already know requires $x > 1$ for convergence and so the function is defined as in Figure E.1. If we simply replace $x \in \mathbb{R}$ with $z \in \mathbb{C}$, we have a continuation to

$$\zeta(z) = \sum_{r=1}^{\infty} \frac{1}{r^z},$$

a complex-valued function of a complex variable. We would expect the complex form to inherit a similar restriction and so it does, as we can see from

$$\sum_{r=1}^{\infty} \left| \frac{1}{r^z} \right| = \sum_{r=1}^{\infty} \left| \frac{1}{e^{z \ln r}} \right|$$

$$= \sum_{r=1}^{\infty} \left| \frac{1}{e^{(\mathrm{Re}(z) + i\, \mathrm{Im}(z)) \ln r}} \right|$$

$$= \sum_{r=1}^{\infty} \left| \frac{1}{e^{\mathrm{Re}(z) \ln r} e^{i\, \mathrm{Im}(z) \ln r}} \right|$$

$$= \sum_{r=1}^{\infty} \left| \frac{1}{e^{\mathrm{Re}(z) \ln r}} \right|$$

$$= \sum_{r=1}^{\infty} \frac{1}{r^{\mathrm{Re}(z)}},$$

which we know converges only for $\mathrm{Re}(z) > 1$. So $\zeta(z)$ makes sense in this domain; pictorially, the shaded region in Figure E.2.

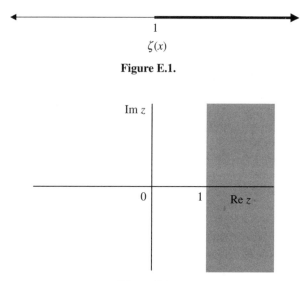

Figure E.1.

Figure E.2.

Euler's product formula remains valid for complex numbers and makes clear that this extended function still has no zeros; so far, this is pretty straightforward stuff. Now for that analytic continuation, which Riemann realized using contour integration.

The complex extension

$$\zeta(z)\Gamma(z) = \int_0^\infty \frac{u^{z-1}}{e^u - 1}\, du$$

of the formula we derived on p. 60, which is valid for $\mathrm{Re}(z) > 1$, suggests a contour integral

$$I(z) = \frac{1}{2\pi i} \oint_{u^-} \frac{u^{z-1}}{e^{-u} - 1}\, du, \quad \mathrm{Re}(z) > 1,$$

for some contour u^-. A useful choice is a path coming from $-\infty$ just below and parallel to the real axis, (semi)circling the origin anticlockwise and returning to $-\infty$ parallel and just above the real axis.

Integrate around C_1, C_2, C_3 separately, therefore putting

$$u = re^{-\pi i}, \qquad u = \rho e^{i\theta}, \qquad u = re^{\pi i},$$

respectively, since on C_3 we are going out to minus infinity (effectively) along the negative imaginary axis, making the argument π; on C_1 we are returning,

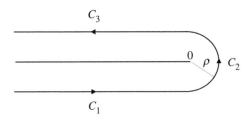

Figure E.3.

making the argument $-\pi$; on C_2 we are traversing a circle of radius ρ. Then

$$2\pi i\, I(z) = -\int_\rho^\infty \frac{r^{z-1}e^{-\pi i z}e^{\pi i}e^{-\pi i}}{e^r - 1}\, dr + \int_{-\pi}^\pi \frac{\rho^{z-1}e^{iz\theta}e^{-i\vartheta}\rho i e^{i\vartheta}}{e^{-\rho e^{i\theta}} - 1}\, d\theta$$
$$+ \int_\rho^\infty \frac{r^{z-1}e^{\pi i z}e^{-\pi i}e^{\pi i}}{e^r - 1}\, dr$$
$$= -e^{-\pi i z}\int_\rho^\infty \frac{r^{z-1}}{e^r - 1}\, dr + \int_{-\pi}^\pi \frac{\rho^z e^{iz\theta} i}{e^{-\rho e^{i\theta}} - 1}\, d\theta$$
$$+ e^{\pi i z}\int_\rho^\infty \frac{r^{z-1}}{e^r - 1}\, dr.$$

So,

$$\pi I(z) = \sin(\pi z)\int_\rho^\infty \frac{r^{z-1}}{e^r - 1}\, dr + \frac{\rho^z}{2}\int_{-\pi}^\pi \frac{e^{iz\theta}}{e^{-\rho e^{i\theta}} - 1}\, d\theta.$$

Taking each integral separately,

$$\left| \frac{\rho^z}{2}\int_{-\pi}^\pi \frac{e^{iz\theta}}{e^{-\rho e^{i\theta}} - 1}\, d\theta \right| = \left| \frac{\rho^z}{2} \right| \left| \int_{-\pi}^\pi \frac{e^{iz\theta}}{e^{-\rho e^{i\theta}} - 1}\, d\theta \right|$$
$$\leqslant \frac{\rho^{\mathrm{Re}(z)}}{2}\int_{-\pi}^\pi \left| \frac{e^{iz\theta}}{e^{-\rho e^{i\theta}} - 1} \right| d\theta$$
$$\leqslant \frac{\rho^{\mathrm{Re}(z)}}{2}\int_{-\pi}^\pi \frac{e^{-\mathrm{Im}(z)\theta}A}{\rho}\, d\theta$$
$$= \frac{\rho^{\mathrm{Re}(z)}}{2}\int_{-\pi}^\pi \left| \frac{e^{iz\theta}}{e^{-\rho e^{i\theta}} - 1}\frac{\rho e^{i\theta}}{\rho e^{i\theta}} \right| d\theta$$
$$= \frac{\rho^{\mathrm{Re}(z)}}{2}\int_{-\pi}^\pi \left| \frac{\rho e^{i\theta}}{e^{-\rho e^{i\theta}} - 1} \right| \left| \frac{e^{iz\theta}}{\rho e^{i\theta}} \right| d\theta$$
$$= \frac{\rho^{\mathrm{Re}(z)}}{2}\int_{-\pi}^\pi \left| \frac{\rho e^{i\theta}}{e^{-\rho e^{i\theta}} - 1} \right| \frac{e^{-\mathrm{Im}(z)\theta}}{\rho}\, d\theta.$$

251

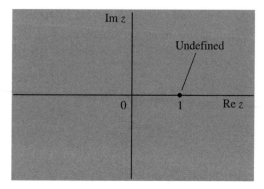

Figure E.4.

But

$$\left| \frac{\rho e^{i\theta}}{e^{-\rho e^{i\theta}} - 1} \right| = \left| \frac{u}{e^{-u} - 1} \right|$$

is bounded for bounded u, let us say by the constant A.

Therefore,

$$\left| \frac{\rho^z}{2} \int_{-\pi}^{\pi} \frac{e^{iz\theta}}{e^{-\rho e^{i\theta}} - 1} d\theta \right| \leq \frac{\rho^{\mathrm{Re}(z)}}{2} \int_{-\pi}^{\pi} A \frac{e^{-\mathrm{Im}(z)\theta}}{\rho} d\theta$$

$$\leq \frac{A\rho^{\mathrm{Re}(z)-1}}{2} 2\pi e^{\pi|\mathrm{Im}(z)|} = \pi A \rho^{\mathrm{Re}(z)-1} e^{\pi|\mathrm{Im}(z)|}$$

and if $\mathrm{Re}(z) > 1$,

$$\frac{\rho^z}{2} \int_{-\pi}^{\pi} \frac{e^{iz\theta}}{e^{-\alpha e^{i\theta}} - 1} d\theta \to 0$$

as $\rho \to 0$ and

$$\pi I(z) = \lim_{\rho \to 0} \sin(\pi z) \int_{\rho}^{\infty} \frac{r^{z-1}}{e^r - 1} dr$$

$$= \sin(\pi z) \int_{0}^{\infty} \frac{r^{z-1}}{e^r - 1} dr = \sin(\pi z) \Gamma(z) \zeta(z)$$

and so

$$\zeta(z) = \frac{\pi I(z)}{\sin(\pi z) \Gamma(z)}$$

and since $\Gamma(z)\Gamma(1-z) = \pi / \sin(\pi z)$ we have that

$$\zeta(z) = \frac{\Gamma(1-z)}{2\pi i} \oint_{u^-} \frac{u^{z-1}}{e^{-u} - 1} du,$$

which is defined and finite for all $z \neq 1$.

The domain of definition is now as in Figure E.4.

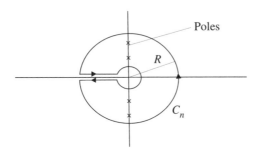

Figure E.5. Outer radius $R = (2N + 1)\pi$.

E.2 ZETA'S FUNCTIONAL RELATIONSHIP

We are going to 'trap' $I(z)$, evaluating it by integrating around a second contour, which in the limit is the same as the one above.

Consider the contour integral

$$I_N(z) = \frac{1}{2\pi i} \int_{C_N} \frac{u^{z-1}}{e^{-u} - 1} \, du,$$

where $\text{Re}(z) < 0$ with the contour shown in Figure E.5 for N a positive integer.

On the outer circle we have $u = Re^{i\theta}$, $-\pi \leqslant \vartheta \leqslant \pi$, and

$$
\left| \frac{u^{z-1}}{e^{-u} - 1} \right| = \left| \frac{(Re^{i\theta})^{z-1}}{e^{-u} - 1} \right|
$$

$$
= \left| \frac{R^{z-1} e^{i\theta(\text{Re}(z)+i\,\text{Im}(z))} e^{-i\theta}}{e^{-u} - 1} \right|
$$

$$
= \left| \frac{R^{\text{Re}(z)-1} R^{i\,\text{Im}(z)} e^{i\theta\,\text{Re}(z)} e^{-\text{Im}(z)}}{e^{-u} - 1} \right|
$$

$$
= e^{-\theta\,\text{Im}(z)} R^{\text{Re}(z)-1} \left| \frac{1}{e^{-u} - 1} \right|
$$

$$
< R^{\text{Re}(z)-1} e^{\pi\,\text{Im}(z)} A < R^{\text{Re}(z)} e^{\pi\,\text{Im}(z)} A
$$

since

$$
\left| \frac{1}{e^{-u} - 1} \right|
$$

is bounded in the region.

So, as N, $R \to \infty$ the contribution to the integral from this part of the contour $\to 0$ and therefore $I_N(z) \to I(z)$.

The function $f(u) = u^{z-1}/(e^{-u} - 1)$ has poles where $e^{-u} - 1 = 0$ and so $u = 2k\pi i$ for $k = 1, 2, \ldots, N$ and for $k = -1, -2, \ldots, -N$ (which is why the outer radius is taken to be $(2N + 1)\pi$). If we are to use Cauchy's Residue

Theorem to evaluate $I_N(z)$, we will need the residues at each of these poles and so we will use the theory of residues to find them:

$$\operatorname*{Res}_{u=2k\pi i} f(u) = \operatorname*{Res}_{u=2k\pi} \frac{p(u)}{q(u)} = \frac{p(2k\pi i)}{q'(2k\pi i)} = \frac{(2k\pi i)^{z-1}}{-1} = -(2k\pi i)^{z-1}.$$

So,

$$I_N(z) = \frac{1}{2\pi i} \int_{C_N} \frac{u^{z-1}}{e^{-u} - 1} \, du$$

$$= -\sum_{k=1}^{N} \{(2k\pi i)^{z-1} + (-2k\pi i)^{z-1}\}$$

$$= -\sum_{k=1}^{N} (2\pi k)^{z-1} (e^{\pi(z-1)i/2} + e^{-\pi(z-1)i/2})$$

$$= -\sum_{r=1}^{N} (2\pi r)^{z-1} 2\cos(\pi(z-1)/2)$$

$$= -2(2\pi)^{z-1} \sin(\pi z/2) \sum_{r=1}^{N} r^{z-1}.$$

Now we recognize that we have integrated around the contour in the opposite direction for $I(z) = \lim_{N\to\infty} I_N(z)$, so we have that

$$I(z) = -\lim_{N\to\infty} I_N(z) = 2(2\pi)^{z-1} \sin(\pi z/2) \sum_{r=1}^{\infty} r^{z-1}$$

$$= 2(2\pi)^{z-1} \sin(\pi z/2)\zeta(1-z)$$

with the convergence guaranteed, as $\operatorname{Re}(1-z) = 1 - \operatorname{Re}(z) > 1$.

Each form of $I(z)$ was established using a different assumption about $\operatorname{Re}(z)$ but the uniqueness of analytic continuation allows this to be disregarded and, combining these two forms for $I(z)$, we get

$$\zeta(z) = \frac{2\pi(2\pi)^{z-1} \sin(\pi z/2)\zeta(1-z)}{\sin(\pi z)\Gamma(z)} = \frac{(2\pi)^z \sin(\pi z/2)\zeta(1-z)}{\sin(\pi z)\Gamma(z)}$$

and we have the promised functional relationship $\zeta(1-z) = \chi(z)\zeta(z)$, where $\sin(\pi z)\Gamma(z)\zeta(z) = (2\pi)^z \sin(\pi z/2)\zeta(1-z)$ for all $z \neq 1$, which becomes $\zeta(1-z) = 2(2\pi)^{-z} \cos(\pi z/2)\Gamma(z)\zeta(z)$.

References

Agar, J. 2001 *Turing and the Universal Machine: The Making of the Modern Computer (Revolutions in Science)*. Icon Books.

Ahlfors, L. V. 1966 *Complex Analysis*. McGraw-Hill.

Aleksandrov, A. D., Kolmogorov, A. N. & Lavrent'ev, M. A. (eds) 1964 *Mathematics, Its Content, Methods and Meaning* (English edn), 3 vols. Cambridge, MA: MIT.

Baillie, R. 1979 Sums of reciprocals of integers missing a given digit. *Am. Math. Mon.* **86**, 372, 374.

Barnett, I. A. 1972 *Elements of Number Theory*. Prindle, Webster & Schmidt.

Barrow, J. D. Chaos in numberland (http://plus.maths.org/issue11/features/cfractions/). Issue 11 of Plus online magazine (http://plus.maths.org).

Baumgart, J. K. (ed.) 1969 *Historical Topics for the Mathematical Classroom*. 31st Yearbook of the National Council of Teachers of Mathematics.

Benford, F. 1938 The law of anomalous numbers. *Proc. Am. Phil. Soc.* **78**, 551ff.

Borwein, J. M. & Borwein, P. B. 1987 The way of all means. *Am. Math. Mon.* **94**, 519–522

Boyer, C. B. & Merzbach, U. C. 1991 *A History of Mathematics*. Wiley.

Browne, M. W. 1998 Following Benford's Law, or looking out for no. 1. *The New York Times* (Tuesday, 4 August).

Burkill, H. 1995/6 G. H. Hardy. *Math. Spectrum* **28**(2), 25–31.

Burrows, B. L. & Talbot, R. F. 1984 Sums of powers of integers. *Am. Math. Mon.* **91**, 394–403.

Burton, D. M. 1980 *Elementary Number Theory*. Allyn & Bacon.

Calinger, R. 2001 *Towards a New Biography of Euler: Historiography*. The Catholic University of America.

Cherwell, Lord 1941 Number of primes and probability considerations. *Nature* **148**, 436.

Cohn, H. 1980 *Advanced Number Theory*. New York: Dover.

Cohn, H. 1971/72 How many prime numbers are there? *Math. Spectrum* **4**(2), 69–71.

Conrey, B. 1989 At least two fifths of the zeros of the Riemann Zeta function are on the critical line. *Bull. JAMS*, pp. 79–81.

Conway, J. H. & Guy, R. K. 1995 *The Book of Numbers*. Copernicus.

Coolidge, J. L. 1963 *The Mathematics of Great Amateurs*. New York: Dover.

Cowen, C. C., Davidson, K. R. & Kaufmann, R. P. 1980 Rearranging the alternating harmonic series. *Am. Math. Mon.* **87**, 17–19.

Davis, P. J. & Hersh, R. 1983 *The Mathematical Experience*. Pelican.

DeTemple, D. W. 1991 The non-integer property of sums of reciprocals of consecutive integers. *Math. Gaz.* **75**, 193–194.

de Visme, G. H. 1961 The density of prime numbers. *Math. Gaz.* **45**, 13–14.

Devlin, K. 1988 *Mathematics, the New Golden Age*. Penguin.

Dowling, J. P. 1989 The Riemann Conjecture. *Math. Mag.* **62**, 197.

Dunham, W. 1999 *Euler: The Master of Us All*. The Mathematical Association of America.

Eves, H. 1965 *An Introduction to the History of Mathematics*. New York: Holt, Rinehart and Winston.

Eves, H. 1969 *In Mathematical Circles: A Two Volume Set*. Kent: PWS.

Eves, H. 1983 *Great Moments in Mathematics Before 1650*. The Mathematical Association of America.

Eves, H. 1983 *Great Moments in Mathematics After 1650*. The Mathematical Association of America.

Eves, H. 1971 *Mathematical Circles Revisited*. Kent: PWS.

Eves, H. 1988 *Return to Mathematical Circles*. Kent: PWS.

Fauvel, J. & Gray, J. (eds) 1987 *The History of Mathematics—A Reader*. Macmillan.

Flegg, G. 1984 *Numbers, Their History and Meaning*. Penguin.

Fletcher, C. R. 1996 Two prime centenaries. *Math. Gaz.* **80**, 476, 484.

Freebury, H. A. 1958 *A History of Mathematics*. Cassell.

Furry, W. H. 1942 Number of primes and probability considerations. *Nature*, **150**, 120–121.

Gardner, M. 1986 *Knotted Doughnuts and Other Mathematical Entertainments*. San Francisco, CA: Freeman.

Glaisher, J. W. L. 1972 On the history of Euler's constant. *Messenger Math.* **1**, 25–30.

Glick, N. 1978 Breaking records and breaking boards. *Am. Math. Mon.* **85**, 2, 26.

Graham, R. L., Knuth, D. & Patashnik, O. 1998 *Concrete Mathematics: A Foundation for Computer Science*. Addison-Wesley.

Gullberg, J. 1997 *Mathematics from the Birth of Numbers*. W. W Norton & Co.

Hardy, G. H. & Wright, E. M. 1938 *The Theory of Numbers*. Oxford.

Hinderer, W. 1993/94 Optimal crossing of a desert. *Math. Spectrum* **26**, 100, 102.

Hodges, A. 1983 *Alan Turing: The Enigma*. Vintage.

Hoffman, P. 1991 *Archimedes' Revenge: The Joys and Perils of Mathematics*. Penguin.

Ingham, A. E. 1995 *The Distribution of Prime Numbers*. Cambridge University Press.

Khinchin, A. I. 1957 *Mathematical Foundations of Information Theory*. New York: Dover.

Kline, M. 1979 *Mathematical Thought from Ancient to Modern Times*. Oxford University Press.

Kôrner, T. W. 1996 *The Pleasures of Counting*. Cambridge University Press.

Kreyszig, E. 1999 *Advanced Engineering Mathematics*. Wiley.

Lagarias, J. C. 2002 An elementary problem equivalent to the Riemann hypothesis. *Am. Math. Mon.*

Le Veque, W. J. 1996 *Fundamentals of Number Theory*. New York: Dover.

Lines, M. E. 1986 *A Number for Your Thoughts: Facts and Speculations About Numbers from Euclid to the Latest Computers*. Adam Hilger.

MacKinnon, N. 1987 Prime number formulae. *Math. Gaz.* **71**, 113–114.

McLean, K. R. 1991 The harmonic hurdler runs again. *Math. Gaz.* **75**, 190, 193.

Maor, E. 1994 *e: The Story of a Number*. Princeton, NJ: Princeton University Press.

Montgomery, H. L. 1979 Zeta zeros on the critical line. *Am. Math. Mon.* **86**, 43–45.

Nahin, P. J. 1998 *An Imaginary Tale. The Story of* $\sqrt{-1}$. Princeton, NJ: Princeton University Press.

Napier, J. 1889 *The Construction of the Wonderful Canon of Logarithms*. Blackwood and Sons.

Newcomb, S. 1881 *Am. J. Math.* **4**, 39–40.

Niven, I. 1961 *Numbers, Rational and Irrational*. Random House.

Olds, C. D. 1963 *Continued Fractions*. The Mathematical Association of America.

Ore, O. 1988 *Number Theory and Its History*. New York: Dover.

Patterson, S. J. 1995 *An Introduction to the Theory of the Riemann Zeta Function*. Cambridge University Press.

Reid, C. 1996 *Hilbert*. Springer.

Rockett, A. M. & Szusz, P. 1992 *Continued Fractions*. World Scientific.

Rose, H. E. 1994 *A Course in Number Theory*. Oxford University Press.

Shannon, C. E. & Weaver, W. 1980 *The Mathematical Theory of Communication*, 8th edn. University of Illinois Press.

Smith, D. E. *A Source Book in Mathematics*. New York: Dover.

Sondheimer, E. & Rogerson, A. 1981 *Numbers and Infinity*. Cambridge University Press.

Spiegel, M. R. 1968 *Mathematical Handbook*. McGraw-Hill. Schaum Series.

Stark, H. M. 1979 *An Introduction to Number Theory*. Cambridge, MA: MIT.

Struik, D. (ed) 1986 *A Source Book in Mathematics 1200 to 1800*. Princeton, NJ: Princeton University Press.

Swetz, F., Fauvel, J., Johansson, B., Katz, V. & Bekken, O. (eds) 1994 *Learn from the Masters (Classroom Resource Material)*. The Mathematical Association of America.

Tall, D. O. 1970 *Functions of a Complex Variable*, vols 1 and 2. New York: Dover.

Wadhwa, A. D. 1975 An interesting subseries of the harmonic series. *Am. Math. Mon.* **82**, 931, 933.

Walthoe, J., Hunt, R. & Pearson, M. 1999 Looking out for number one (http://plus.maths.org/issue9/features/benford/). Issue 9 of Plus online magazine (http://plus.maths.org).

Webb, J. 2000 In perfect harmony (http://plus.maths.org/issue12/features/harmonic/). Issue 12 of Plus online magazine (http://plus.maths.org).

Weisstein, E. 2002 *The CRC Concise Encylopedia of Mathematics*, 2nd edn. Chapman & Hall/CRC, London.

Wells, D. 1986 *The Penguin Book of Curious and Interesting Numbers*. Penguin.

Wilf, H. S. 1987 A greeting and a view of Riemann's Hypothesis. *Am. Math. Mon.* **94**, 3, 6.

Willans, C. P. 1964 A formula for the nth prime number. *Math. Gaz.* **48**, 413, 415.

Wright, E. M. 1961 A functional equation in the heuristic theory of primes. *Math. Gaz.* **45**, 15–16.

Young, R. M. 1991 Euler's constant. *Math. Gaz.* **75**, 187, 190.

Related Web Resources

Clay Mathematics Institute (www.claymath.org).

MacTutor History of Mathematics Archive (www-groups.dcs.st-and.ac.uk/~history).

Math Archives (http://archives.math.utk.edu/topics/history.html).

On-line Encyclopedia of Integer Sequences (www.research.att.com/~njas/sequences).

Name Index

Subject Index

alternating Zeta function, 206
analytic continuation, 247
  idea of, 191
Analytical Engine, 84
anthyphairetic ratio, 93
Apery's constant, 42
arithmetic mean
  compared with other means, 119
  of logs of numbers, 8
  of partial quotients, 159
Astronomia Nova, 11
average deficits of quotients, 113
average distance of a planet from the Sun,
  12
average number of divisors, 112

Babylonian counting system, 147
Babylonian identity, 1, 119
Babylonian tablets, 3
Basel Problem, 38
Benford's Law, 145
Bernoulli Numbers, 41, 79, 81, 83, 86, 194
Bernoulli's integral, 44
Bertrand Conjecture
  statement and use with H_n, 25
  use for an upper bound for p_n, 167
Bessel Equation, 107
Big Oh Notation, 219
birthday paradox, 147
Bletchely Park, 213
Blue Pig, 214
Bohr–Mollerup Theorem, 56
Briggsian logarithms, 10
British Imperial system, 149
Brun's constant, 30

calculus of residues, 245
Cauchy–Riemann equations, 227
Chebychev weighted prime counting
  function, 183
collecting a complete set, 130
Complement Formula, 59
complex differentiation, 225
complex integration, 232

complex logarithms, 231
conditional convergence, 102
Constructio, 3
continued fraction
  appearance in an interview problem,
    137
  connection with Pell's equation, 97
  definition of, 93
  form of special numbers, 96
  of an approximation to gamma, 105
  result giving the minimum size of the
    denominator of a rational Gamma,
    97
  statistical behaviour of, 155
  three results of, 95
  use in geometric harmony, 122
  use in musical harmony, 123
convergents, 94
cosine integral, 106
Cossist, 81
critical strip, 195
crossing the desert, 127
cumulative density function, 151

deficits of quotients, 113
derivative of $\pi(x)$, 199
Descriptio, 3
  Preface, 3
difference between 22/7 and π, 96
Digamma function, 58
digital analysis, 155
divergence of the prime harmonic series, 62
Dowling's poem, 216

Enigma Code, 213
entropy, 139
equivalence of PNT and the size of the xth
  prime, 182
Eratosthenes
  sieve of, 171
error function, 106
estimating $\pi(x)$, 164
Euclidean prime, 28